Nanoscale Science and Engineering Forum 2021

Held at the 2021 AIChE Annual Meeting

Boston, Massachusetts, USA and Online
7 - 11 November and 15 - 19 November 2021

ISBN: 978-1-7138-5715-0

Printed from e-media with permission by:

Curran Associates, Inc.
57 Morehouse Lane
Red Hook, NY 12571

Some format issues inherent in the e-media version may also appear in this print version.

Printed with permission by Curran Associates, Inc. (2022)

For permission requests, please contact AIChE
at the address below.

AIChE
120 Wall Street, FL 23
New York, NY 10005-4020

Phone: (800) 242-4363
Fax: (203) 775-5177

www.aiche.org

Additional copies of this publication are available from:

Curran Associates, Inc.
57 Morehouse Lane
Red Hook, NY 12571 USA
Phone: 845-758-0400
Fax: 845-758-2633
Email: curran@proceedings.com
Web: www.proceedings.com

TABLE OF CONTENTS

633089 Graduate Student Award Session: Development of Stable Targeted Nano-, Encapsulated Manganese Oxide (NEMO) Particles for Early Breast Cancer Diagnosis By MRI

Abstract:

The ultimate goal of this research project is to develop a novel contrast agent to reduce the number of false-positive and false-negative imaging results associated with current breast cancer imaging techniques. Gadolinium chelates are the current clinically approved contrast agents for breast magnetic resonance imaging (MRI), but they are always "on" and highlight any vascularized structure. Due to their lack of targeting, both benign and malignant breast tumors are enhanced with gadolinium, resulting in high false-positive rates up to 25%. By utilizing tumor-targeted pH-sensitive manganese oxide (MnO) nanoparticles, a contrast agent can be developed that will only turn "on" after internalization into cancer cells and dissolution within low pH endosomes/lysosomes.

Nanocrystalline MnO was synthesized through thermal decomposition of Mn(II) acetylacetonate in dibenzyl ether and oleylamine at 280°C. After confirmation of nanocrystalline MnO size and chemistry with TEM, XRD, and FTIR, MnO nanocrystals were encapsulated within poly(lactic-co-glycolic acid) (PLGA) and poly(ethylene glycol) (PEG) through single emulsion to form Nano-, Encapsulated, Manganese Oxide (NEMO) particles. Particles with different PEG percentages (2.5, 5, and 10% w/w) and PLGA terminations (carboxylic acid vs. ester) were synthesized and incubated in water over 24 hours to assess particle stability. The hydrodynamic diameter and charge of NEMO particles were measured pre- and post-incubation with dynamic light scattering (DLS) and zeta potential, respectively. For carboxylic acid terminated PLGA, a higher PEG % resulted in enhanced particle stability due to a more negative zeta potential that favored particle-particle repulsion (2.5% PEG: -12.89 mV; 5% PEG: -14.31 mV; 10% PEG: -16.33 mV). Surprisingly, increased PEG % had the opposite effect with ester terminated PLGA; particles became less stable with higher PEG due to a less negative zeta potential that reduced interparticle repulsion (2.5% PEG: -16.51 mV; 5% PEG: -12.61 mV; 10% PEG: -8.557 mV). As higher PEG density can reduce nanoparticle cell uptake, NEMO particles with 2.5% PEG ester terminated PLGA were chosen as the optimal particle for further experiments.

To enable specific uptake by cancer cells, a tumor-targeting peptide against underglycosylated mucin-1 (uMUC-1) was attached through click chemistry using copper (I) as the catalyzer. The uMUC-1 targeting peptide conjugated NEMO particles were characterized with DLS and a fluorometric assay to assess hydrodynamic size (~180 nm) and correct targeting attachment, respectively. For NEMO particles to cause a positive T_1 signal on MRI, MnO must dissolve and release Mn^{2+} at low pH. To evaluate the Mn^{2+} release and MRI properties, NEMO particles were incubated at 3 different pHs over time including pH 7.4 (blood pH), 6.5 (pH of tumor extracellular space), and 5 (endosome/lysosome pH), and supernatants were evaluated using inductively coupled plasma-optical emission spectroscopy (ICP-OES) and 1T MRI. NEMO particles released the most Mn^{2+} at pH 5 (~93% at 24 hr), which produced the lowest T_1 value with the brightest MRI signal at the first two time points (848 ms at 1 hr, and 886 ms at 2hr, pH 5). The substantial MRI contrast generated after just 1-2 hours indicates that NEMO particles can result in a measurable contrast change within clinically relevant time frames.

To assess *in vivo* stability and toxicity, different dosages of tumor-targeted NEMO particles (15, 7.5, 3, and 1.5 mg Mn/kg) were injected once a week for three weeks into BALB/c female mice. Weight, behavior, and appearance were scored daily to measure toxic effects. Only the highest dosage, 15 mg Mn/kg, had an impact on mouse weight during the first 24 hr (~1 g weight loss), and resulted in a slight decrease in natural and provoked behavior; mice regained weight and normal behavior within 2 to 3 days post-injection. Based on these results, overall NEMO particles were well tolerated *in vivo* over 3 weeks. In future studies, 7.5 mg Mn/kg will be utilized to maximize Mn dose for enhanced MRI signal while avoiding the weight and behavioral effects associated with 15 mg Mn/kg. Future work includes further *in vitro* studies of NEMO particle cell labeling and toxicity, along with *in vivo* MRI of tumor-bearing mice to assess the specific detection of malignant versus benign breast tumors with NEMO particles.

625102 Multispectral Fingerprinting Resolves Dynamics of Nanomaterial Trafficking in Primary Endothelial Cells

Abstract:

Intracellular vesicle trafficking involves a complex series of biological pathways used to sort, recycle, and degrade extracellular components, including engineered nanomaterials (ENMs) which gain cellular entry *via* active endocytic processes. A recent emphasis on routes of ENM uptake has established key physicochemical properties which direct certain mechanisms, yet relatively few studies have identified their effect on intracellular trafficking processes past entry and initial subcellular localization. Here, we developed and applied an approach where single-walled carbon nanotubes (SWCNTs) play a dual role - that of an ENM undergoing intracellular processing, in addition to functioning as the signal transduction element reporting these events in individual cells with single organelle resolution. We used the unique optical properties exhibited by non-covalent hybrids of single-stranded DNA and SWCNTs (DNA-SWCNTs) to report the progression of intracellular processing events *via* two orthogonal hyperspectral imaging approaches of near-infrared (NIR) fluorescence and resonance Raman scattering. A positive correlation between fluorescence and G-band intensities was uncovered within single cells, while exciton energy transfer and eventual aggregation of DNA-SWCNTs were observed to scale with increasing time after internalization. These were confirmed to be consequences of intracellular processes using pharmacological inhibitors of endosomal maturation, which suppressed spectral changes through two distinct mechanisms. An analysis pipeline was developed to colocalize and deconvolute the fluorescence and Raman spectra of subcellular regions of interest (ROIs), allowing for single-chirality component spectra to be obtained with sub-micron spatial resolution. This approach uncovered a complex relationship between DNA-SWCNT concentration, fluorescence intensity, environmental transformations, and irreversible aggregation resulting from the temporal evolution of trafficking events. Finally, a spectral clustering analysis was applied to delineate the dynamic sequence of processes into four distinct populations, allowing stages of the intracellular trafficking process to be identified by the multispectral fingerprint of encapsulated DNA-SWCNTs.

625198 Continuous Free-Flow Isoelectric Separation of Extracellular RNA Nanocarriers from Plasma

Abstract:

Extracellular RNA (exRNA) nanocarriers secreted in biological fluids play a crucial role in a variety of complex cellular functions by trafficking signaling molecules from cell to cell. exRNAs abundant in physiological fluid (such as blood, urine, lymph fluids) are encased and protected by three classes of nanocarriers: Extracellular vesicle (EVs), Lipoprotein (LLPs) and Ribonucleic protein (RNPs). Several recent studies show that exRNA nanocarriers from cancer tumor cells contain a unique set of molecular cargos that can indicate the different stage of tumor progression and hence has a potential to serve as diagnostic and prognostic biomarkers. It is thus very important to separate these carriers and analyze them separately. The currently available separation techniques use density-gradient multiple stage ultracentrifugation, immunocapture and ultrafiltration techniques, which are cumbersome and personnel intensive. Moreover, owing to overlapping nanocarrier size distributions and densities, these isolation technologies often suffer from low yield, poor purity, and quantification inaccuracy. To address these shortcomings, we take the advantage of their distinct isoelectric points and developed a high-throughput membrane-based free-flow isoelectric separation technology for rapid isolation of exRNA carriers (exosome, LLPs and RNPs) with high yield and purity within minutes from a plasma sample.

The device is configured by a single upstream channel, wedged between two bipolar membranes, further connected to the downstream channel (Fig. 1). Under reverse bias, bipolar membrane depletes the ions along their field lines to reduce the ionic current and increase the field strength due to the ion depletion zone at the junction. The excessively high field at the junction of the bipolar membrane allows splitting of water into H+ and OH-. The generated H+ and OH-, then precisely driven out from their respective ion exchange membrane. Further, their diffusion and continuous injection in the downstream channel allows the establishment of the pH gradient from 2 to 11, with distinct pH bands. A constant transverse electric field up to $\pm$150 V/cm is then applied in the downstream channel for the separation of three carriers in three isolated chambers.

The activated device is successfully used for the generation of a constant pH gradient for isoelectric focusing and monitoring of separation (Fig. 2). Free-Flow isoelectric focusing is demonstrated with the fluorescently tagged high-density lipoprotein, low-density lipoprotein and ribonucleoprotein protein, covering a range of isoelectric point from 4 to 8. Separation of carriers was accomplished within a residence time of 10 seconds at a flow rate of 3ml/hr. Furthermore, the characterization of outlet sample by UV spectrometry and Gel-Electrophoresis suggests the purity of the isolated particles more than 90%.

The defined microfluidic device has shown the potential and benefits of providing a high-throughput and high-precision separation of disease cargos with different range of pIs, within a few seconds. The continuous free flow allows to process a large volume of sample in a short time. Collectively, this device enables the sophisticated approach for early disease detection and may provide an effective technique to validate and identify new disease biomarkers in body fluids. With a proper extension, it can also be used to purify nanoparticles like virus vaccines, exosome drug carriers, amyloid-beta aggregates, peptide assemblies etc.

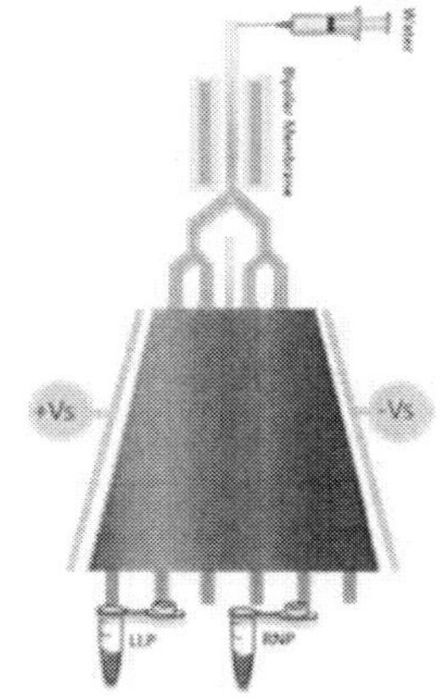

Fig 1: Schematic of Isoelectric focusing device.

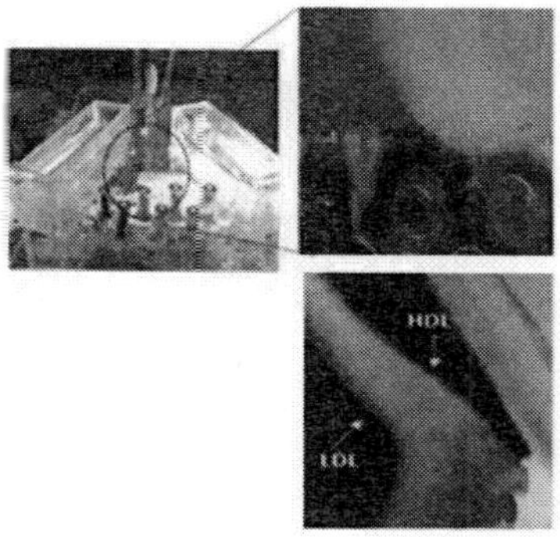

Fig 2: Experimental image of pH gradient and separation of lipoproteins.

625150 Superparamagnetic Nanobead and Electrodeposited Magnetic Nanoporous Membrane for Immunocapture of Specific Exosomes

Abstract:

Extracellular vehicles (EVs) are cell-derived small membrane-bounded vesicles containing various functional cargos. Both their protein and miRNA contents are suggested to be potential cancer biomarkers. Moreover, by profiling RNA cargo inside EVs with specific surface proteins, better statistics could be obtained for tumor source, size and type. Traditional immunocapture, like precipitation, antibody functionalized columns, and functionalized microbeads, has issues of low yield due to probe saturation. Their throughputs are also limited by the long incubation time (8-48 hours) because of the long transport and reaction time. A solution facilitating the yield and throughput is to use nanomagnetic beads. The large surface area per volume gives it more binding sites. The smaller size leads to higher diffusivity and shorter incubation time (~30 minutes). However, it is hard to trap due to its superparamagnetic nature. The magnetic force on them is proportional to the gradient of field squared rather than to the field. For the commonly used small magnets with low field penetration, the attraction is only effective within one radius of the curvature of the sharp tip at the magnet.

We report an electrodeposited magnetic nanoporous membrane device for high-efficiency recovery of immunocapture nanobeads to overcome this key obstacle in nanobead immunoassay. A layer of NiFe is electroplated on a track-etched PET membrane with a gold seed layer. Magnets are put on both sides of the membrane to magnetize the membrane. By flowing the solution through the membrane, the nanobeads are brought into the region of intense magnetic fields and field gradients at the edge of the nanopore. Thus, we can achieve a high recovery rate of nanobeads and captured EVs. More than 99% of beads can be recovered from the solution with a throughput of 3mL/hr on a single device. The uncaptured EVs can go through the straight pores and be collected in the flow-through. We proved our device's efficiency and specificity using HDL (high density lipoproteins) as a model - more than 80% of HDL is recovered, nearly doubled from the commercial kits. The high and consistent yield of our system provides quantification potential for biological studies. We further demonstrated our system's applications in exosome capturing and profiling, EGFR-positive (Epidermal Growth Factor Receptor-positive) EVs characterization, and purification of HDL-enriched EV samples.

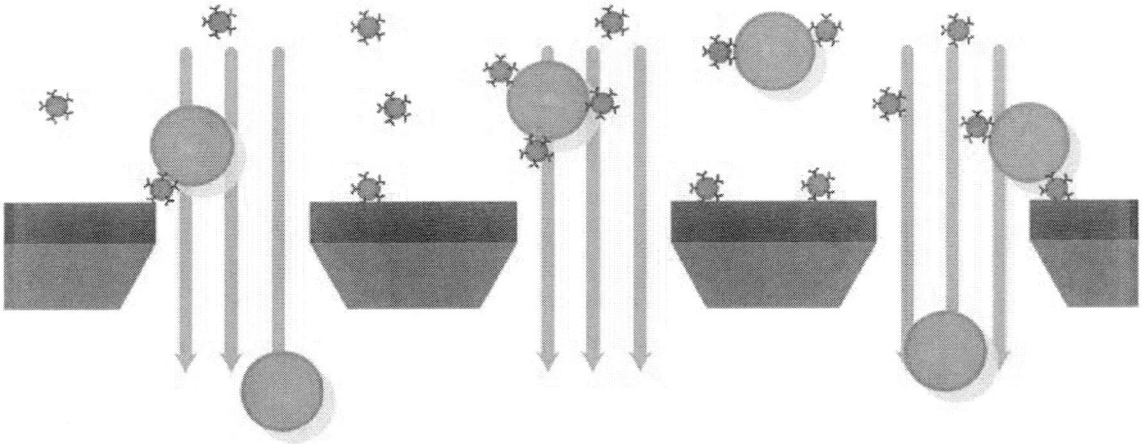

632323 Dynamics Monitoring of Temozolomide and 5-Aminoimidazole-4-Carboxamide for Glioblastoma Using Fluorescent Nanosensors

Abstract:

The efficacy of chemotherapeutics is often affected by the timing, quantity and frequency of dosages. Real-time, dynamic measures of potency may supplement or in some cases replace reliance on bio-imaging. Towards this end, in this work we develop synthetic molecular recognition sites for temozolomide (TMZ), and its decomposition product 5-aminoimidazole-4-carboxamide (AIC), grafting them onto fluorescent nanoparticles capable of forming optode or other biosensor interfaces to monitor drug efficacy in real-time. Near-infrared fluorescent single-walled carbon nanotubes, wrapped using specific DNA oligonucleotides, are encapsulated within biocompatible, poly(ethylene glycol) diacrylate hydrogels and enable the selective recognition of an anti-cancer drug, TMZ, on U-87 MG human glioblastoma cells. The sensors exhibit a fluorescence response to TMZ with a detection limit of 30 mM and used to track the progression of glioblastoma death following TMZ administration. For *in vivo* demonstration, the sensors were able to monitor TMZ dynamics and flux in SKH-1E mice. In fetal pig brain, deeply implanted sensors detected AIC dynamics by employing wavelength-induced frequency filtering technique. These results enable real-time detection of chemotherapeutic concentration and metabolism *in vivo* with potential to increase the efficacy of cancer treatments.

622351 Data Management Schema Design for Effective Nanoparticle Formulation for Probing and Treating Neurological Disease

Abstract:

Background: Nanomedicine has limited translation from pre-clinical research to clinical application for non-cancer neurological disease [1]. To improve translation, it is important to understand the dynamic interactions between nanomedicine platforms and the complex brain environment. To investigate dynamic nanoparticle-brain interactions with a range of nano-based delivery systems, it is useful to develop a database that can organize and integrate key experimental variables and factors that might influence the effectiveness of nanotherapeutics in the brain.

Database management systems already commonly exist in clinical medical applications. For example, The Brain Database is a clinical neuroanatomy database developed over three decades ago [2]. However, pre-clinical data management schemas have different complexities than clinical applications. Pre-clinical databases need management schemas that logically connect experimental methodologies, which can be highly repetitive, vary in quality, are prone to rapid iteration and evolution, and often research lab or facility-specific. There is limited or non-existent literature on effective database management for nanoformulations used preclinically, but similar approaches for utilizing databases exist in computational cell biology for managing data-rich biological cell process data [3]. For example, BioNumbers is a computational cell biology database that stores and organizes key quantitative features from cell biology [4]. The BioNumbers database is composed mainly via natural language processing of the literature, containing one specific number per property. In contrast, nanoformulation-brain databases are focused on organizing wet lab experimental research prior to the publication phase and often have repetitive experiments with varying values for each property. A successful standardized data management schema for pre-clinical nanotherapeutic experimental pipelines has three main effects: (1) increased sustainability for data organization and storage, (2) increased insight into related variables and their interactions, and (3) increased searchability of specific results or methods.

Methods:

To show the efficacy of a well-designed pre-clinical database management schema for nanoformulation data, we create a database using PostgreSQL that performs three functions: (1) Obtains the nanoparticle formulation variables that produce specific nanoparticle sizes and charges; (2) Connects nanoformulations to all related experimental elements, including nanoformulation methodology, biological specimen prep, and biological specimen characterization; and (3) Defines a standard lab protocol for regularly updating the database as new nanoformulation experiments are carried out.

Our first step was to design the entity-relationship diagram to relate three main aspects: experiment, nanoformulation, and biological specimen (e.g. application). We then created a process flow diagram as a visual representation of data management from the chemical engineering, experimentalist perspective (Figure 1). This process flow diagram was converted into an entity-relationship diagram using LucidChart. With LucidChart, we were able to visualize our relationship details within the database, including all variables, keys, and variable types. From LucidChart, we directly exported the database schema in PostgreSQL and subsequently imported the data onto a local server. We hosted the database locally due to the overall data size (51.5 kilobytes). Before we uploaded all of our data to the schema, we cleaned and standardized the data sets, which were collected from nine independent researchers.

After the database was created, we developed standard queries for accessing nanoformulation and biological data based on commonly used variables. For the nanoformulation data, we use three key physical characteristics of nanoformulations to inform our queries: size, zeta potential, and polydispersity index (PDI). Nanoparticle size and surface charge can influence nanoparticle passage across the blood-brain barrier and penetrate within the brain parenchyma [5]. Zeta potential is an indirect measure of nanoparticle surface charge and PDI is an indicator of particle size uniformity. We developed standardized queries which allow us to search biological characterization information including catalase activity, cytotoxicity over time, and glutathione levels for experiments that use nanoformulations in biological specimen. Finally, we developed a standardized method of inputting data into a cloud-based database by creating group spreadsheets with commonly used variables for each nanoformulation technique and characterization method. We additionally standardized a method for assigning unique keys for each individual nanoformulation, researchers, characterization experiment, and biological specimen based on input order and time information.

Results:

Once the database was set up, data input, and queries written, we evaluated our data management schema's effectiveness via two methods. The first approach evaluated the times it took for each query to run and return results. The first set of our queries selects all nanoparticle formulations with specific characterization features in either size, zeta potential, polydispersity, or a mix of these three variables. The second set of our queries obtains all nanoparticle formulations and their formulation variables for specific nanoparticle features for either single emulsion, double emulsion, or nanoprecipitation methodologies. The slowest query ran with an average time of 72.5 milliseconds. Before creating the database, to obtain a list of nanoformulation methodologies and their variables for a specific nanoformulation characterization result, an interested researcher would need to make personal asks to other researchers or perform additional nanoformulation experiments. Both options are time-consuming and ineffective. With a standard database, it is possible for researchers to easily access previous experiments from other researchers in an efficient way by searching directly for a variable of interest.

The second evaluation method visualizes our nanoformulations' physical characteristics for each query. The visualization of our nanoformulation data is important because it allows us to check that the queries are doing what we intended and provides valuable scientific insight into our formulation data. For example, if we query the size, zeta potential, and PDI, it is clear that the queries successfully minimize our data set to those within user-defined parameters. Our first query restricted the size of formulated nanoparticles to within 50 to 100 nanometers. This query restricted the total results from 538 formulations to 237. Our second query added an additional restriction on the zeta potential values to reside between -10 and 10 millivolts. Query 2 restricted our results from 237 formulations to 160. Finally, we restricted our overall formulations with an additional parameter on PDI values between 0.0 and 0.2 which brought our viable nanoformulation with these parameters to 109. These evaluation plots also provide unique insight into how our different nanoformulation methods compare to one another. For example, we can now query our data for specific nanoparticle feature ranges and then compare our single emulsion methodologies to our double emulsion methodologies. We are also able to gain insight from our zeta potential chart as is. From queries 2 and 3, we learned that although our nanoformulations are within the neutral range (-10 to 10 mV), they still tend towards negative values. The database works to successfully relate formulations and methodologies with specific formulation outcomes.

Conclusions:

Developing and utilizing a database management schema on already established experimental pipelines provided new insights, effective data storage, and data querying options. We were able to query our nanoformulations to obtain approximately 100 nanoformulation batches that fit the requirements imposed in this project. We also have shown a proof-of-concept that our experimental data can benefit from a well-structured database management schema. Additionally, we are currently expanding the database to match nanoformulation experiments to biological experiments. We are doing this by creating additional tables for biological specimen variables and related biological characterization methods. For example, an *in vivo* rat experiment can be connected to behavioral testing results, litter information, and any tissue processing. Once biological information is input into the database, we use an experiment tag to connect nanoformulations and their characterizations to the biological specimen they were tested in. The database organizes these connections for efficient querying, and then we develop queries for nanoformulations based on biological features such as species, age, or behavioral results.

Finally, we plan to develop standard practices for discussing, planning, and participating in better data management. These practices will lead to better data utilization and data longevity, and greater reproducibility and interpretability of datasets generated from independent researchers. Insights into relationships and trends in nanomedicine formulations have benefitted from developing this database management schema. A standard data management schema for experimental nanoformulation experiments decreases the need for repetitive experimentation, connects nanoformulation variables with biological outcome in an interpretable and efficient way, and enables science-informed querying of any standardized variables. The overall benefits of a standardized database contribute to efficient experimentation and cross-experimental insight, which will ultimately improve pre-clinical to clinical translation of nanoformulations.

References:

[1] Anselmo, A.C. and S. Mitragotri, *Nanoparticles in the clinic: An update.* Bioengineering & Translational Medicine, 2019. **4**(3).

[2] Wetheim, S., *The Brain Database: A Multimedia Neuroscience Database for Research and Teaching.* 1989.

[3] Loew, L.M. and J.C. Schaff, *The Virtual Cell: a software environment for computational cell biology.* Trends in Biotechnology, 2001. **19**(10): p. 401-406.

[4] Milo, R., et al., *BioNumbers—the database of key numbers in molecular and cell biology.* Nucleic Acids Research, 2010. **38**(suppl_1): p. D750-D753.

[5] Helmbrecht, H., et al., *Governing transport principles for nanotherapeutic application in the brain.* Current Opinion in Chemical Engineering, 2020. **30**: p. 112-119.

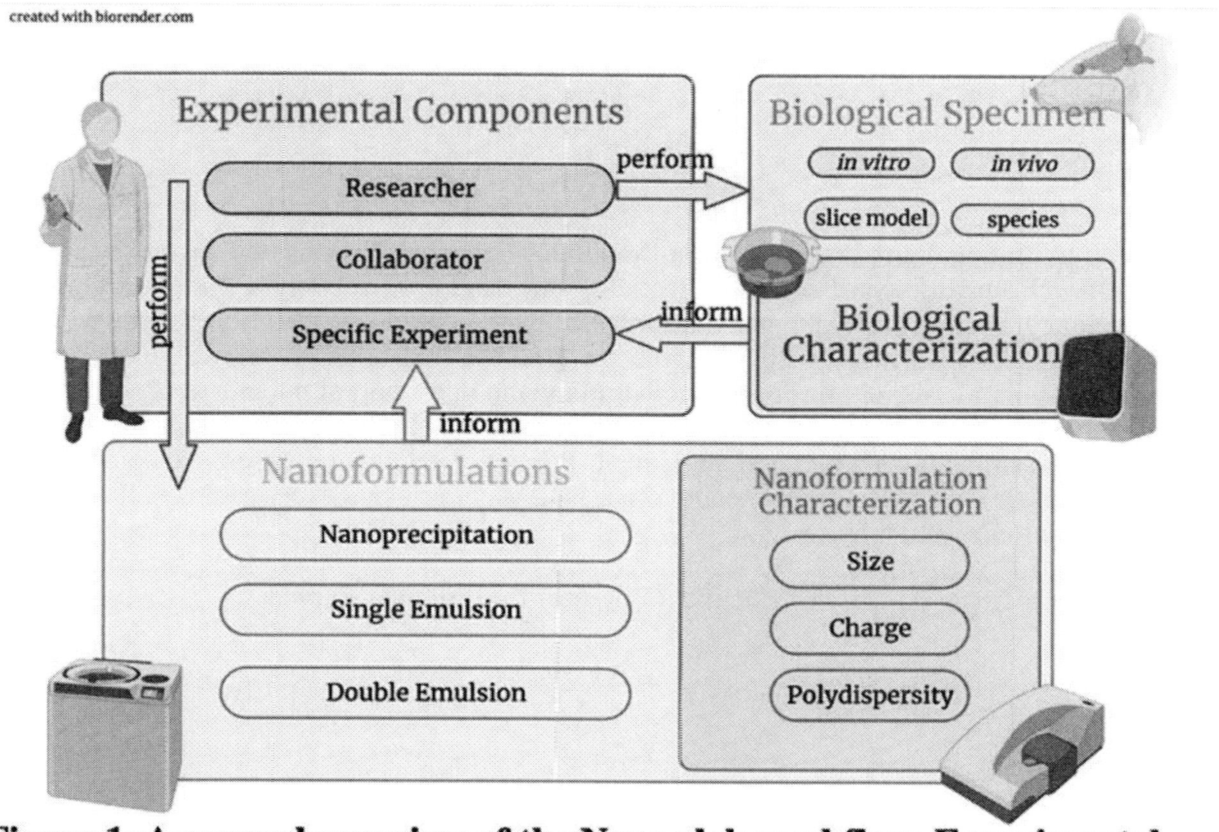

Figure 1. A general overview of the Nance lab workflow. Experimental components include specific experiment details, researcher, and collaborator details. Biological details include the specimen: *in vitro, in vivo,* slice model, and species alongside the characterization of these specimens. The nanoformulations include the methodology and characterization of resulting nanoparticles.

624106 Stimuli Responsive Self-Folding 3D Graphene Architectures

Abstract:

Due to their exceptional properties and ultra-thin dimensions, two-dimensional layered materials (2DLMs) have been widely utilized in electronics, optics and drug delivery. Prior research has focused on integration with rigid planar substrates which can limit applicability. Here, we describe monolayer graphene integration with a flexible and gradient crosslinked epoxy polymer SU8. We demonstrate reversible solvent responsive self-folding and unfolding of integrated photodetector devices using these flexible substrates. Experiments are guided by finite element mechanics and electromagnetic simulations with good agreement. We anticipate that this approach could be used to create novel 3D devices using graphene and other 2DLMs of relevance to wearables, flexible electronics, photonics and biosensors.

626188 Rheoelectric Characterization of Oxidized Carbon Nanoparticles As Slurry Active Materials

Abstract:

Promising battery systems for renewable energy sources are ones that will provide resilient, flexible power while reducing the carbon emissions, oil demands, and recurring costs of fossil fuel. To this end, the electrochemical flow capacitor (EFC) is an innovative energy storage technology comprised of a slurry active material stored externally from the electrochemical flow cell. This spatial decoupling absent from conventional batteries allows low-cost scaling of EFCs as the price of the overall package asymptotes towards slurry raw material costs. Energy-dense slurries require electrically conductive particles to form a charge-carrying network in an electrolyte solution. Previous work has focused on carbon black because it is inexpensive, has a high surface-area-to-volume ratio, and has excellent electrical conductivity. Unfortunately, carbon black suspensions under flow become undesirably viscous, leading to unacceptable pumping costs for EFC operations. Mitigating high viscosities, however, comes at a trade-off with improving energy storage efficiency as the slurry electrical conductivity and specific capacitance also increase with carbon black content, as described by electrosorption. We hypothesize that by engineering the nanoscale forces of carbon black within suspension, we can design suspension electrodes with better rheological and electrochemical performance. Carbon black particles were functionalized by acid oxidation, presenting a surface chemistry that exhibits a strong, repulsive surface charge and confers colloidal stability. Changes to the size and the surface charge density of the primary agglomerates were derived from dynamic light scattering and zeta potential measurements, respectively. Chemical composition of the surface was obtained through X-ray photoelectron spectroscopy, complemented with Fourier-transform infrared spectroscopy. The active surface was further characterized with porosity and BET surface area measurements through gas sorption. Simultaneous shear stress and impedance measurements of homogenized Carbon black slurries with varying salt concentrations were acquired from a custom-built, cup and bob flow cell. We will develop structure-property relationships for several Carbon black slurries by linking flow-induced microstructural evolutions to observed rheoelectric performances. We anticipate the presence of an optimal position between low viscosity and high conductivity that maximizes the performance to cost ratio of carbon black slurry electrodes.

628887 Controllable Synthesis of Hybrid Nanocomposite Structures *Via* Laser Ablation Technique for Electrochemical Energy Storage and Conversion Devices

Abstract:

Hybrid carbon-based nanocomposites fabricated from Metal-Organic Frameworks (MOFs) comprised of transition metals are promising candidates as Oxygen Reduction Reaction (ORR) electrocatalysts due to their superior electron conductivity and high porosity. To this end, a considerable number of studies have focused their attention on enhancing the structural, chemical, and physical properties of such hybrid materials by optimizing their size, morphology, and structure while compositing them with other materials. Our recent works have shown that Laser Ablation Synthesis in Solution (LASiS) techniques can provide a fast, economical, and environmentally friendly method for fabricating MOF-based functional hybrid nanocomposites (HNCs). Our objective in this study is two-fold: 1) to evaluate the efficacy of this technique to rationally engineer the size and shape of MOF structures through modifying solution-phase (reagent concentration and temperature), laser (output power and ablation time), and post-treatment (pyrolysis) parameters, and 2) examine the electrocatalytic performance, stability, and economic feasibility of non-precious metals based HNCs for ORR. Specifically, herein we have optimized the LASiS technique with post-pyrolytic treatments to synthesize HNCs in carbonaceous matrix derived from composite Co-Co_3O_4@ZIF-67 nanostructures that exhibit significantly enhanced ORR electrocatalytic performance.

628771 Fundamental Process Parameters and Fiber Formation Mechanisms in Wet-Spinning of Carbon Nanotube Fibers

Abstract:

The mechanical and electrical properties of neat carbon nanotube fiber (CNTF) produced from solutions of pristine CNTs in chlorosulfonic acid are rapidly improving. CNTF is on track to be the strongest synthetic fiber by the end of the decade, and it already surpasses high performance carbon fibers and polymeric fibers in flexibility and electrical/thermal conductivities. As properties begin to exceed those of widely adopted materials (polyaramids, polypropylene, polyethylene, etc.), process cost and environmental efficiency need to be competitive to bring the technology to market. While some research efforts have focused on the impact of the CNT material on CNTF properties, many details of the processing conditions that lead to optimal CNTF properties have not been reported, nor have the mechanisms which limit the formation of CNTF under different conditions. It is understood that CNT solution concentration and spin-line throughput are two primary efficiency determining variables, but the impact of these variables on the spinning window (the range of parameters over which certain grades of CNTF are able to be produced) for the wet spinning process have not yet been thoroughly studied. Here, we seek to define and understand the experimental spinning windows to identify which process parameters govern optimal CNTF properties. We studied the impact of extrusion and take-up parameters, CNT solution concentration, and coagulation conditions on the CNTF properties to determine viable spinning windows. To gain insight and understanding to correlate rheological properties with spinning windows, we measured the viscosity and shear rate behavior of CNT solutions using capillary rheometry. Additionally, we used in-line microscopy on the spinning system to visualize the phenomena occurring at the exit of the spinneret, including die swell, coagulation rate, and mechanisms of filament formation and stability. Finally, we identify the most important factors for CNTF properties and spinning efficiency, and we propose a path forward for efficient production of high-quality CNTF.

631744 Investigating the Mechanism of Temperature-Induced Graphene Self-Folding in Water

Abstract:

The popularity of graphene origami - eyed for applications in electronics, energy harvesting, molecular storage and membrane development - has been on the rise for the last few years, with the world's tiniest microchips created through the same process just in 2021. With hopes of making graphene origami easier, temperature-induced self-folding of graphene using thermosensitive polymers like poly(N-isopropylacrylamide) (PNIPAM) has been demonstrated experimentally. This process, attributed to the coil-to-globule transition of PNIPAM, is however, poorly understood, making the creation and control of these self-folded three dimensional (3D) structures difficult. For this reason, we employ coarse-grained (CG) molecular dynamics (MD) simulations to develop an accurate molecular-level mechanism for the self-folding of PNIPAM-grafted graphene in presence of an explicit solvent. Our analyses of contacts, energetics and structure of PNIPAM-solvent interface indicate that the process is actually initiated by the weakening of PNIPAM-water interactions at higher temperatures, resulting in strong hydrophobic graphene-PNIPAM interactions. Densely grafted PNIPAM chains above LCST are observed to stabilize through chain aggregation and a coil-to-globule transition may not always be observed. By further varying system parameters like graphene width and polymer coverage, we show that graphene folding only occurs when the total energy of the final conformation is lower than the total energy of the initial conformation, indicating the stabilization of the structure after folding. We also find that PNIPAM-PNIPAM interactions compete with graphene-PNIPAM interactions and folding occurs when the latter dominates. Through the molecular-level insights obtained in this study, we thus provide energy criteria for graphene-folding that can be essential for controlling and tuning this process *via* temperature control.

631072 Utilizing Unique Thermodynamic Equilibria of Co-Surfactant States Around Nanotubes for Optical Biosensors

Abstract:

The unique optical properties of single walled carbon nanotubes (SWCNTs) make them a suitable candidate for biosensing applications. Although difficulty in separating monochiral fractions has limited their use as fluorescent sensors, surfactants offer a convenient and scalable approach towards suspending nanotubes in solvents. Surfactant structure around the nanotube plays an important role in stability and separation of (n,m) species. The presence of thermodynamically stable co-surfactants states enables selective coating of species in solution. Our group has previously utilized co-surfactant solution approach to elute (6,5) species at a very specific ratio of sodium dodecyl sulphate (SDS)/sodium deoxycholate (DOC). This idea was used to selectively target other chiralities and obtain high concentration monochiral in the nanotube solution.

Here, we present a new method of utilizing this thermodynamic equilibria for selectively eluting a mixture of chiralities and attaching ligands on them without separating monochiral fractions. Adsorption of added surfactant leads to reorganization of the surfactant shell around nanotube which leads to solvatochromatic shifts that can be probed through optical spectroscopy. Difference in surfactant binding strength can be used to protect chiralities against further coating using addition of controlled ratios of co-surfactant solutions to nanotubes. We investigated the role of covalent and non-covalent functionalization to the nanotube surface for obtaining best biosensing responses. This led to study the role of different ligands for functionalizing different chiralities for simultaneous detection of analytes in solution. This approach provides a unique way of using SWCNTs as biosensors by selectively functionalizing different chiralities without obtaining monochiral fractions.

627633 Gold Nanoparticles Deposited inside Hierarchical Zeolites Catalyze Diverse Groups of Aromatic Alcohols

Abstract:

The slow internal mass transport, induced by long diffusion path length, of reactants and corresponding products to and from the active site reduces transport efficiency of single-sized zeolite catalysts. Comparatively, hierarchical zeolites consist of two (or more) level of highly interconnected porosity (microporosity, mesoporosity, and macroporosity), which enhance the mass transport and molecular accessibility in catalytic reactions. Our previous work indicated that the oxidation of benzyl alcohol, catalyzed by the gold nanoparticles inside the mesopores of SBA-15 zeolites, could be potentially limited by internal mass transport due to the long and tortuous diffusion path of mesoporous zeolites. The goal of this work is to design flexible catalyst synthesis technique to tune catalysts' performance. Herein, we demonstrate a catalyst synthesis technique, directly depositing SwiS-stabilized gold nanoparticles inside hierarchical zeolites to alleviate the internal mass transport limitation in zeolite-based catalysts. Previously, we had demonstrated the direct deposition of switchable surfactant (SwiS)-stabilized gold nanoparticles inside the mesoporous SBA-15 zeolites. We hypothesize that SwiS-stabilized gold nanoparticles can be directly diffused inside both micropores and mesopores of hierarchical zeolites via incipient wetness technique, and the synthesized material can catalyze the oxidation of diverse groups of aromatic alcohols. We used transmission electron microscopy (TEM) to verify whether the deposition is successful. Based upon the knowledge from our previous work of depositing gold nanoparticles inside mesoporous zeolites, we anticipated the deposition was successful. Next, we investigated if this deposition affected the size of gold nanoparticles. Our previous work had shown that SwiS-stabilized gold nanoparticles well maintain size during deposition. Hence, we expected the size of gold nanoparticles to be well preserved. To test the second half our hypothesis, we tested the catalytic activities of the synthesized material in the oxidation of aromatic alcohols with different molecular size. Comparatively, the catalytic activities of gold nanoparticles inside mesoporous zeolite were tested in the same reaction conditions. We expected the gold nanoparticles inside the hierarchical zeolites to catalyze more types of aromatic alcohol than those inside mesoporous zeolites in oxidation reaction.

629922 Development of Ion-Conducting Polymers for Hydrogen Electrochemical Energy Conversion Technologies

Abstract:

Ion-conducting polymers (often called ionomers) are used as polymer electrolyte membranes as well as catalyst binders in membrane electrode assemblies, are a key component of electrochemical energy conversion and storage technologies such as fuel cells, electrolyzers, and flow batteries. The use of ion-conducting polymers in these clean energy technologies has a long history and there has been considerable progress in developing low-cost, high-performance ion-conducting polymers in recent years. In this presentation, we discuss some highlights from recent progress at the Bae group of Rensselaer Polytechnic Institute in the development of advanced ion-conducting (cationic and anionic) polymers, their state-of-the-art performance in electrochemical energy conversion devices (hydrogen fuel cells, electrolyzer, redox flow battery) and technology transfer activity to commercialize materials for real-world applications.

628713 MOF-Derived PtCo/Co_3O_4 nanocomposites in Carbonaceous Matrices As High-Performance ORR Electrocatalysts Synthesized *Via* laser Ablation Techniques

Abstract:

Rapidly expanding global energy demands due to fast-paced human-technology interfaces have propelled fuel cell technology as a sustainable energy-conversion alternative. Nonetheless, the rational development of such technology demands the engineering of low-cost durable materials that can exhibit and retain specific electrocatalytic activities to sustain the electrochemical energy conversions. Over the years, our group has extensively developed the now-patented laser ablation synthesis in solution-galvanic replacement reaction (LASiS-GRR) technique as a non-equilibrium, one-pot and green route for synthesizing disparate heteronanostructured intermetallic composites[1,2] including Metal Organic Frameworks (MOF)[3] without using any surfactants/ligands, that can be tuned for diverse interfacial activities (energetics, optical, electrochemical). Herein, we specifically present LASiS-GRR as a facile route for the synthesis of Pt-Co bimetallic nanoparticles (NPs) encapsulated in MOF phases of ZIF-67 that are concurrently formed during the laser ablation process.[4] Upon pyrolytic post-treatments, these composite structures are found to be partially altered, giving rise to the formation of graphitic shell coated Pt-Co bimetallic nanocomposites (NCs) that are spatially framed onto Co_3O_4-decorated carbonaceous matrices. Our investigations into the effects of Pt precursor (K_2PtCl_4) concentrations during the LASiS-GRR technique indicate that increasing K_2PtCl_4 initial concentrations in the solution-phase facilitate a higher degree of Pt-Co alloying. Subsequent electrochemical characterizations reveal the ability of these hybrid NCs in carbonaceous matrices to exhibit superior electrocatalytic activities towards oxygen reduction reaction (ORR) while maintaining long-term stability in highly concentrated alkaline media (KOH 1M). Specifically, the NCs synthesized from LASiS under 125 $mg \cdot L^{-1}$ K_2PtCl_4 solution, demonstrate an extraordinary 5-fold enhancement in their specific mass activities as compared to state-of-the-art commercial Pt catalysts. These outstanding performances are attributed to the unique configurations arising from cooperative catalytic activities (spill-over effects) between the graphitic shell coated bimetallic Pt-Co nanoparticles and the electrochemically active Co_3O_4-decorated carbon matrix support.

References:

1. Hu S, Goenaga G, Melton C, Zawodzinski TA, Mukherjee D. PtCo/CoOx nanocomposites: Bifunctional electrocatalysts for oxygen reduction and evolution reactions synthesized via tandem laser ablation synthesis in solution-galvanic replacement reactions. *Applied Catalysis B: Environmental.* 2016;182:286-296.
2. Hu S, Tian M, Ribeiro EL, Duscher G, Mukherjee D. Tandem laser ablation synthesis in solution-galvanic replacement reaction (LASiS-GRR) for the production of PtCo nanoalloys as oxygen reduction electrocatalysts. *Journal of Power Sources.* 2016;306:413-423.
3. Ribeiro EL, Davari SA, Hu S, Mukherjee D, Khomami B. Laser-induced synthesis of ZIF-67: a facile approach for the fabrication of crystalline MOFs with tailored size and geometry. *Materials Chemistry Frontiers.* 2019;3(7):1302-1309.
4. Ribeiro EL, Davis EM, Mokhtarnejad M, Hu S, Mukherjee D, Khomami B. MOF-derived PtCo/Co3O4 nanocomposites in carbonaceous matrices as high-performance ORR

electrocatalysts synthesized via laser ablation techniques. *Catalysis Science & Technology*. 2021.

624956 Strong *sp-D* Orbital Hybridization Driven Pt-Graphene Hybrid Catalysts for Direct CO_2 Hydrogenation to Formic Acid

Abstract:

Formic acid (FA) is one of the most promising compounds for hydrogen storage for its high hydrogen density, non-toxicity, and great stability under ambient conditions, which can be made from CO_2. Although the noble metal or alloy catalysts have validated their outstanding activity to the dehydrogenation of FA into H_2 and CO_2, the hydrogenation of CO_2 towards FA is relatively difficult due to the nature of relatively weak CO_2 adsorption compared with HCOO adsorption on the metal surfaces. Indeed, thermal CO_2 hydrogenation to FA has been a challenge for a decade since thermal CO_2 hydrogenation is endothermic process. Therefore, the efficient transformation of CO_2 into renewable energy sources such as FA is a crucial for further development of new energy technology. In this study, we investigate the CO_2 hydrogenation towards FA at ambient temperature on a new Pt-Graphene hybrid catalyst using DFT method. This new Pt-Graphene hybrid catalyst consists of Pt monolayer and graphene via hybridizing *sp* and *d* orbitals between carbon and Pt atoms. The sp-d hybridization significantly reduces the local density of states of *d*-band and at the same time increases *s*-band near the Fermi level which leads to the strong chemisorption of CO_2 at the Pt surface. From our DFT calculations, it is discovered that the CO_2 hybridization towards FA is exothermic reaction on Pt-Graphene catalyst. Furthermore, the microkinetic modeling is implemented to identify overall turnover frequencies. By analyzing various metal-Graphene hybrid catalysts, we will also address physical and chemical intuition for the design of new hybrid catalysts.

Computational Design of Active Hybrid Interface Energy Materials from Scratch and Data Science

Hoje Chun, Seungjun Hong, Minjoon Hong, Byungchan Han
Department of Chemical & Biomolecular Engineering, Yonsei University, Seoul 03722, Korea (Republic of)

Abstract

Hybrid-epitaxial nanomaterials have been played a key role in wide range of electrochemical energy devices such as Li-air batteries, photovoltaic devices, and fuel cells. Innovative design for highly functional nanoparticles has been, however, delayed. In molecular level computational electrochemistry new research paradigm has been established, which substantially incorporates IT-based artificial intelligence (AI) technology into machine learning algorithm. Using the new computational methodology high-throughput screening of promising nanoparticle candidates has been attempted for various desired applications.

Whether the frontier approach is successful or not is significantly controlled by the reliability and accuracy of input database. It is true that substantial amounts of the data are come by previous literatures and often ab-initio calculations with idealized model systems. The conditions in which the data were generated may be so different from the *operando* circumstances of the target materials. To secure extreme-level integrity of the database the in-situ measurement of nanoparticle structures should be carried out, from which the reliable correlation of the structure-performance-design principle can be identified.

Using first-principles calculations we studied nanoparticles with adsorbate ligands in liquid solution to establish three-dimensional (3D) structure and property database, which are, then, analyzed through AI-based neural-network approach with high speed and accuracy. The information includes sizes, lattice distortions, and defects with picometer resolution under non-vacuum conditions. The computational outcomes are rigorously validated from the 3D liquid-cell electron microscopy. The approach is indeed 'knowledge-based' AI, which can be expected to make groundbreaking ways toward the quantum nanoarchitecture for hybrid interface materials.

628887 Controllable Synthesis of Hybrid Nanocomposite Structures *Via* Laser Ablation Technique for Electrochemical Energy Storage and Conversion Devices

Abstract:

Hybrid carbon-based nanocomposites fabricated from Metal-Organic Frameworks (MOFs) comprised of transition metals are promising candidates as Oxygen Reduction Reaction (ORR) electrocatalysts due to their superior electron conductivity and high porosity. To this end, a considerable number of studies have focused their attention on enhancing the structural, chemical, and physical properties of such hybrid materials by optimizing their size, morphology, and structure while compositing them with other materials. Our recent works have shown that Laser Ablation Synthesis in Solution (LASiS) techniques can provide a fast, economical, and environmentally friendly method for fabricating MOF-based functional hybrid nanocomposites (HNCs). Our objective in this study is two-fold: 1) to evaluate the efficacy of this technique to rationally engineer the size and shape of MOF structures through modifying solution-phase (reagent concentration and temperature), laser (output power and ablation time), and post-treatment (pyrolysis) parameters, and 2) examine the electrocatalytic performance, stability, and economic feasibility of non-precious metals based HNCs for ORR. Specifically, herein we have optimized the LASiS technique with post-pyrolytic treatments to synthesize HNCs in carbonaceous matrix derived from composite Co-Co_3O_4@ZIF-67 nanostructures that exhibit significantly enhanced ORR electrocatalytic performance.

632721 Enhancing the Performance of Mxene Supercapacitor in Ionic Liquid By Expanding Interlayer Spacing

Abstract:

In the last decade, MXenes have become a fast-growing family of electrically conductive two-dimensional transition metal carbides and nitrides. Benefitting from their unique layered structures, MXenes are highly conductive and exhibit the capability to host various organic and inorganic molecules. Supercapacitors using MXenes as electrode materials have achieved high specific capacitances at high scan rates in aqueous solutions.[1] Nonetheless, the limited electrochemical window of aqueous electrolytes remains an obstacle for maximizing the energy density of MXene supercapacitors. In order to produce supercapacitors that have comparable energy densities with conventional batteries, widening the potential window is an effective approach considering the specific energy equation for capacitive systems. Ionic liquids (ILs) have been widely used as electrolytes in supercapacitors due to wide electrochemical windows and excellent chemical stabilities However, ILs are typically comprised of large, bulky ions which limit the accessibility and transportation in porous materials.

Experiments have shown that pre-intercalation of organic ions (specifically, tetraalkyl ammonium cations) with varying chain lengths into MXenes can widen the interlayer spacing (d-spacing) up to ~1.5 nm, which results in significantly increased capacitances.[2] In this work, classical molecular dynamics (CMD) simulations were used to analyze the capacitive performance of these pre-intercalated MXenes at a molecular-level and try to find the underlying charge storage mechanism. In these CMD simulations, the constant potential method (CPM) was used to apply partial charges on MXene atoms and maintain the potential at the MXene surfaces. In comparison with the constant charge method, the CPM is able to more accurately describe the charge distribution at solid surfaces. CPM is particularly useful for IL-MXene systems which consist of solid surfaces of varied functional groups and ions with complicated structures.

Through the use of CMD, systems of varying d-spacings and surface groups are considered. Structural properties of ILs within the MXene layers are analyzed to compute capacitive properties, whereas the dynamic properties are described through self-diffusion coefficients and the time evolution of the total charge on the MXene surfaces. By analyzing both capacitive and dynamic performances, we hope to find a guideline for simultaneously improving the energy density and power density of supercapacitors.

Reference

1. Lukatskaya, M. R. *et al.* Ultra-High-Rate Pseudocapacitive Energy Storage in Two-Dimensional Transition Metal Carbides. *Nat. Energy* **2**, 17105 (2017).
2. Liang, K. *et al.* Engineering the Interlayer Spacing by Pre-Intercalation for High Performance Supercapacitor MXene Electrodes in Room Temperature Ionic Liquid. Submitted (2021).

623742 Thin Films of Metal Organic Frameworks Interfaces Obtained By Laser Evaporation

Abstract:

Properties such as large surface area, pore volume, high chemical and thermal stability, and structural flexibility render zeolitic imidazolate frameworks (ZIFs) as well-suited materials for gas separation, chemical sensors, and optical and electrical devices. However, for such applications, film processing is a prerequisite.

Several methods demonstrated controlled film formation and relied on pulsed-laser deposition (PLD), atomic layer deposition (ALD), molecular layer deposition (MLD), liquid-phase epitaxy (LPE), chemical solution deposition (CSD) or Langmuir-Blodgett layer-by-layer (LBL) deposition, just to name a few. However, such methods failed to demonstrate that resulting films were uniform in nature. Moreover, several of these techniques required long reaction times and were not able to provide user-controlled, consistent morphologies or crystallinities for the deposited films. Furthermore, PLD for instance, is known to lead to high degradation of some of the films because of the vapor-related condition encountered during such specimen's growth. With above listed limitations, developing the next generation of controlled deposition means capable to ensure maximum heterogeneity as well as continuous, highly oriented film formation, with control over film functionality and integrability, is urgently needed.

Herein, Matrix Assisted Pulsed Laser Evaporation (MAPLE) was successfully used as a single-step deposition process to fabricate ZIF-8 films. MAPLE versatility was previously demonstrated during the deposition of magnetite silica, organic targets, SnO_2 colloidal, TiO_2 nanoparticle, kanamycin functionalized magnetite nanoparticles and polyaniline, just to name a few. Using MAPLE was expected to reduce the logistical burden associated with ZIF-8 thin film deposition which is made as brittle crystals or insoluble powders that are not amenable to common surface-processing techniques. By correlating laser fluency and controlling the specific transfer of lab-synthesized ZIF-8, films with user-controlled physical and chemical properties were obtained. Films' characteristics were evaluated by Scanning Electron Microscopy (SEM), energy-dispersive X-ray (EDX) spectroscopy, X-ray diffraction (XRD), Fourier Transform Infrared (FTIR) and X-ray photoelectron (XPS) spectroscopies, respectively. Analysis showed that frameworks of ZIF-8 can be deposited successfully and controllably using the gentle MAPLE technique to yield to polycrystalline films on selective substrates. Deposited films maintained the integrity of individual ZIF-8 framework, while undergoing minor crystalline and surface chemistry changes. No significant changes in particle size homogeneity and uniformity were however observed. Our study demonstrated control over both the MAPLE deposition conditions and the outcome, as well as the suitability of the deposition method to create composite architectures. We envision that controlled deposition of MOF films using MAPLE could be tuned to create structures with application in sensors or membrane or to form seeding layers in which such deposited films serve as nucleation centers for growth of catalytic precursors. Such tuning capability combined with the ability to deposit on a variety of substrates from metals to oxides, or polymers, all at room temperature, could allow for increased versatility of deposited thin films while largely maintaining their physico-chemical characteristics.

627626 *High-Throughput Sonochemical Synthesis of Nanocrystals in Deep Eutectic Solvents*

Abstract:

Semiconductor quantum dots (QD) have seen increasing interest in the past decade as these nanocrystals have tunable electronic/optical properties, while retaining structural stability. They have found application in catalysis, biomedicine, photovoltaic and light-emitting diodes (LED) technologies. One route for the synthesis of these materials is through the use of sonochemistry, in which an ultrasound wave is applied to a liquid sample containing dissolved nanocrystal precursors. An appealing choice for solvents are deep eutectic solvents (DESs): a class of materials with varied applications including catalysis, synthesis, extraction processes, drug solubilization and battery electrolytes. Their appeal for applications in nanocrystal synthesis stems from their high surface tension, which allows for aggressive cavitation, low vapor pressure, desirable for avoiding the formation of aerosol during the sonication step, and the tunability of their bulk properties. Additionally, it is also possible to embed the QD precursors in the solvent itself, as either a type I DES or a type III. A breadth of candidates with varying concentrations can be used to form DESs, leading to a vast design space. A liquid handling robot, a high-throughput sonication apparatus and a plate reader were used to test the formulation of different DES candidates and QD precursors. Through the use of high throughput experiments and data-driven design strategies we aimed to accelerate the optimization of materials based on their physicochemical and photoluminescence properties.

630739 Synthesis of Egyptian Blue and Mechanisms

Abstract:

Calcium copper silicate ($CaCuSi_4O_{10}$), known as Egyptian Blue (EB), is a pigment that exhibits a strong near-infrared (NIR) (910 nm) fluorescent emission when exposed to a strong visible light source. This property makes it a novel candidate for use as a basis for nanomaterial-based sensors in surface water, as its IR signal is very strong in comparison with current commonly used IR reporters. Also, IR signals penetrate biological samples better, and are scattered less than visible range signals. Since the goal of the research is to produce a uniform coating of Egyptian Blue nanoparticles, first the synthesis must be controlled to obtain a uniform morphology. The reaction conditions for Egyptian Blue synthesis were explored with thermogravimetric analysis coupled with mass spectrometry, and high temperature X-ray diffraction (HT-XRD) measurements in order to better understand the reaction mechanism. A reproducible maximum in EB yield was observed around 7 °C/min in the HT-XRD analyses. Here we report the reaction conditions, the yields, and the photoluminescence characterization of the synthesized EB samples.

632099 Impact of Mesoporous Silica-Encapsulated Gold Core-Shell Nanoparticle Structure on Solvent-Free Aerobic Benzyl Alcohol Oxidation

Abstract:

In industrial settings, there is a consistent demand for heterogeneous catalysts that are both active and selective. Previous research has shown that encapsulation of gold nanoparticles within sufficiently thin mesoporous silica shells allowed for active, selective oxidation of benzyl alcohol. Importantly, it was found that mesoporous silica shells do not inhibit diffusion of reactants to the gold surface, due to the nanoscale diffusion path. External research has similarly reported a mesopore structure improving catalytic activity by promoting beneficial reactant orientation, despite obstructing the active surface. To demonstrate this effect is present in core-shell nanoparticles, a method for consistently modifying silica shell thickness without impacting other catalyst properties was developed.

Shell thickness was found to increase consistently as a function solution alkalinity. Increasing the concentration of sodium hydroxide also resulted in larger gold core diameters, making controlled studies difficult. To solve this problem, core diameter and shell thickness were decoupled by separating the synthesis into two phases. Surfactant-stabilized gold nanoparticles were first synthesized using a reverse microemulsion synthesis method, commonly used to produce monodisperse gold nanoparticles in the 7 to 10 nm range. These nanoparticles were then added to the synthesis reactor, in place of the unmixed reagents that originally formed the stabilized nanoparticles in situ.

Core-shell nanoparticles with increasing shell thicknesses were used to catalyze solvent-free aerobic benzyl alcohol oxidation. As predicted, nanoparticles with thicker silica shells demonstrated higher catalytic activity on a calculated surface atom basis. In order to compare selectivity, reaction times were extended such that all catalysts reached the same conversion benchmark. Catalysts of all shell thicknesses reached roughly equivalent selectivity towards benzaldehyde, suggesting that pore length is not directly responsible for catalyst selectivity.

632777 Quantitative Spatial Distribution of Rare Earth Dopants in $NaGdF_4$ Nanoparticles

Abstract:

Exploring distribution of rare earth dopants within nanoparticles are crucial to understand the relationship between dopants and their optical properties, and thusly to rationally improve and optimize their optical properties. Energy dispersive X-ray spectroscopy (EDS) element mapping is the well accepted method to map out the distribution of the dopants within the nanoparticles. Firstly, EDS line mapping provides a two dimensional and depth-averaged qualitative distribution of the dopants in the nanoparticles and, secondly, the mapping is only done on few selected nanoparticles which might not represent the distribution within all the nanoparticles. Here, we employ anomalous small angle X-ray scattering (ASAXS) measurements to quantitatively measure the three-dimensional distribution of $Yb^{3+}/Y^{3+}/Lu^{3+}$ in $NaGdF_4$. These measurements revealed a radial gradient distribution of dopants ($Yb^{3+}/Y^{3+}/Lu^{3+}$) from the core to the outer edges of the $NaGdF_4$ nanoparticles. Moreover, shells with the same dopant species and concentration were epitaxially grown on Yb^{3+}-doped sample ($NaGdF_4$:18%Yb,2%Er, denoted as Gd:Yb) to synthesis core/shell (Gd:Yb@Gd:Yb) and core-shell-shell (Gd:Yb@Gd:Yb@Gd:Yb) samples. We discovered heterogeneous distribution of ytterbium ions within the shell-regions. The overall features of the distribution of the sensitizer ions in the host matrix obtained from detailed ASAXS analyses were also qualitatively confirmed by independent EDS measurements. These findings are useful for rationalizing shell-dependent emission properties, understanding the emission mechanisms in complex core/shell nanostructures, and for creating models for multifunctionality and optimal performance in applications.

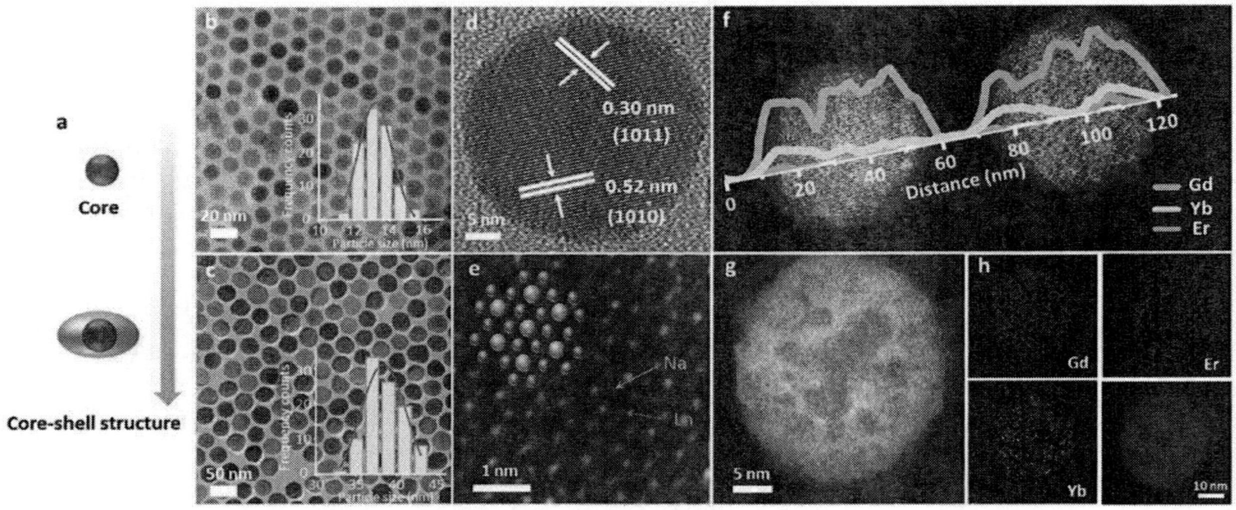

627633 Gold Nanoparticles Deposited inside Hierarchical Zeolites Catalyze Diverse Groups of Aromatic Alcohols

Abstract:

The slow internal mass transport, induced by long diffusion path length, of reactants and corresponding products to and from the active site reduces transport efficiency of single-sized zeolite catalysts. Comparatively, hierarchical zeolites consist of two (or more) level of highly interconnected porosity (microporosity, mesoporosity, and macroporosity), which enhance the mass transport and molecular accessibility in catalytic reactions. Our previous work indicated that the oxidation of benzyl alcohol, catalyzed by the gold nanoparticles inside the mesopores of SBA-15 zeolites, could be potentially limited by internal mass transport due to the long and tortuous diffusion path of mesoporous zeolites. The goal of this work is to design flexible catalyst synthesis technique to tune catalysts' performance. Herein, we demonstrate a catalyst synthesis technique, directly depositing SwiS-stabilized gold nanoparticles inside hierarchical zeolites to alleviate the internal mass transport limitation in zeolite-based catalysts. Previously, we had demonstrated the direct deposition of switchable surfactant (SwiS)-stabilized gold nanoparticles inside the mesoporous SBA-15 zeolites. We hypothesize that SwiS-stabilized gold nanoparticles can be directly diffused inside both micropores and mesopores of hierarchical zeolites via incipient wetness technique, and the synthesized material can catalyze the oxidation of diverse groups of aromatic alcohols. We used transmission electron microscopy (TEM) to verify whether the deposition is successful. Based upon the knowledge from our previous work of depositing gold nanoparticles inside mesoporous zeolites, we anticipated the deposition was successful. Next, we investigated if this deposition affected the size of gold nanoparticles. Our previous work had shown that SwiS-stabilized gold nanoparticles well maintain size during deposition. Hence, we expected the size of gold nanoparticles to be well preserved. To test the second half our hypothesis, we tested the catalytic activities of the synthesized material in the oxidation of aromatic alcohols with different molecular size. Comparatively, the catalytic activities of gold nanoparticles inside mesoporous zeolite were tested in the same reaction conditions. We expected the gold nanoparticles inside the hierarchical zeolites to catalyze more types of aromatic alcohol than those inside mesoporous zeolites in oxidation reaction.

COVID-19: Prevent Infection, Diagnosis and Vaccination using Nanotechnology

Areeg M Dabbish[1], Mahmoud Saeed Mahmoud[2], Ahmed S. Mahmoud[3], Robert W. Peters[4], and Mohamed K. Mostafa[5]

[1]Department of biology, School of Science and Engineering, The American University in Cairo, New Cairo, Egypt

[2]Al-Azhar University Faculty of Medicine, Damietta, Egypt

[3]Scientific Research Development unit, Egyptian Russian University (ERU), Badr, Egypt

[4]Department of Civil, Construction and Environmental Engineering, University of Alabama at Birmingham, Birmingham, AL, USA

[5]Faculty of Engineering and Technology, Badr University in Cairo (BUC), Cairo, Egypt

Graphical abstract

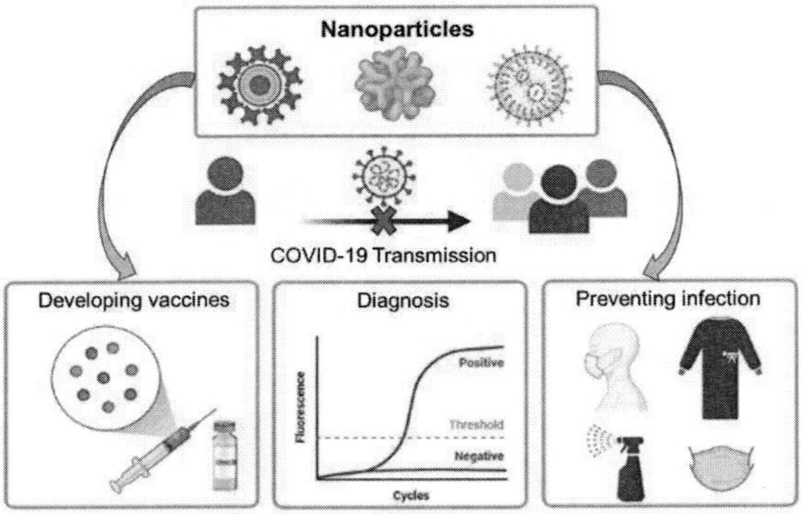

Abstract

The world has been struggling with the major health crisis of the coronavirus disease 2019 (COVID-19) pandemic since December 2019. The highly virulent nature and the mutation rates of the novel virus resulted in less sensitive detection techniques. The main objective of this review paper is to present the possibility of using nanotechnology for COVID-19 prevention, diagnosis, and for developing vaccination. There are different research activities conducted using dissimilar types of masks to control COVID-19 transmission, and their results showed that using single fabric mask with holes was not effective for filtering coronavirus 2019, however, the charged polyvinylidene difluoride (PVDF) nanofiber prevent fast-spreading of COVID-19 and the filtration efficiency may reach 98%. Many researchers suggested that nanosorbents can be used to adsorb COVID-19 virus depending on size, surface area, and the presence of appropriate functional groups. The color of nanosorbents changes relatively by increasing the amount of sorbent virus from the air (virus concentration) and can be measured using different spectral techniques. Nanotechnology can also be an ideal solution to overcome the drawbacks of inactivated vaccines. The common traditional vaccines which depend on live-attenuated or inactivated pathogens present different risks from reflection to pathogenic virulence or guide to a weak immune response. So, the non-traditional vaccines based on nanoparticles (NPs) can enhance strong immunogenicity and antigen presentation by controlling the surface, shape, size and functional groups.

Keywords: COVID-19; Diagnosis; Nanotechnology; Prevent infection; SARS-CoV-2; Vaccination.

1. Introduction

The global problem of coronavirus disease 2019 (COVID-19) pandemic; caused by the novel severe acute respiratory syndrome coronavirus 2 (SARS-CoV-2) has a negative impact on the public health, owing to its ability to spread and mutate over a short period of time (Tu et al., 2020; Zeroual et al., 2020). So, it is necessary to prevent any other environmental infections to control virus separation (Mahmoud et al., 2021; Mostafa & Peters, 2016). The highly virulent nature of the novel virus urges the necessity of exploring rapid and highly sensitive detection approaches for COVID-19. One of the critical challenges facing the diagnosis of COVID-19 is sharing common symptoms with other respiratory disorders. Additionally, positive COVID-19 tests were recorded for some asymptomatic patients. Thus, the molecular detection of the viral genetic material or viral particles is an essential step in the diagnosis (Ramakrishnan et al., 2021). Generally, quantitative real time polymerase chain reaction (RT-qPCR) is the most widely used technique for SARS-CoV-2 screening (Abdeldayem et al., 2022). However, one of the major drawbacks of RT-qPCR is the possibility of human errors that could occur during the viral RNA extraction prior to genome quantifications. Such non-automated procedure results in time consumption, technical errors and less sensitive outcomes. Consequently, alternative automated diagnostic techniques with higher accuracy were urgently needed to restrain the spread of the SARS-CoV-2 during the outbreak (Somvanshi et al., 2020).

SARS-CoV-2 is an enveloped, positive single-stranded RNA virus. The viral particles are characterized by spherical to pleomorphic shape and size ranges from 60 to 140 nm. Viral RNA is tightly coiled, composed of 29811 nucleotide bases, coated with structural and non-structural proteins. The major structural proteins involve spike (S), envelop (E), matrix (M) and nucleocapsid (N) proteins (Fig. 1.). Spike proteins are organized as homotrimers, emerged from the viral lipid envelop forming the significant corona-like structure. Spike proteins are also essential for initiation of SARS-CoV-2 entry into host cell through binding to the angiotensin-converting enzyme 2 (ACE2) receptor on the cell's surface. Any modification in the structural arrangement and in the spatial configuration of these components contributes to loss of functionality and thus viral inactivation (Cascella et al., 2021; Dabbish et al., 2021; H et al., 2020). Based on the nanometric dimensions of SARS-CoV-2 with its nanostructure core-shell, approaches involving nanotechnology open promising horizon in the early and sensitive detection of SARS-CoV-2. The newly emerged techniques have revealed versatile insights in the diagnosis and treatment of several infectious diseases including COVID-19 relying on the use nanoscale materials to target the viral infection cycle (Ruiz-Hitzky et al., 2020) .

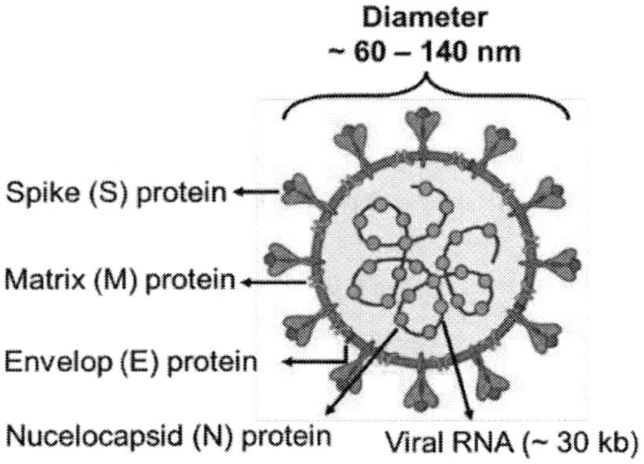

Figure 1. SARS-CoV-2 structure.

2. Diagnosis

In the context of diagnosis, nanotechnology was explored for its ability to develop novel diagnostic techniques for SARS-CoV-2 and to improve strategies of the detection in order to minimize the burden on the healthcare systems.

2.1. Nanoparticles in viral detection

Nanoparticles (NPs) possess several distinctive properties including nano-scaled particle size, improved solubility, increased half-life and lower adverse effects, which ameliorate the product surface adaptability and hasten viral targeting (Rangayasami et al., 2021; Tang et al., 2021) (Hamdy et al., 2019). Previous studies were conducted with advanced nanotechnologies techniques to detect and study the behavior of viral infection using chemically prepared NPs, quantum dots and natural nano-biomaterial (KK, 2005; W et al., 2013). Others suggested the use of nanosorbent to adsorb SARS-CoV-2 depending on size, surface area, and the presence of functional groups. The color of nanosorbent is changed relatively by increasing the amount of sorbent virus from the air (viral concentration) and can be measured by using different optical techniques (N et al., 2020). The use of gold nanoparticles (AuNPs) in the detection of viruses was previously studied based on the facts that AuNPs is capable of binding with some proteins, resulting in surface plasmon resonance (SPR) that can be measured in visible light (Draz & Shafiee, 2018). AuNPs have shown an ability to inhibit amyloid-beta (Aβ) fibrillations which responsible for some neurodegenerative diseases, throughout the interaction between amyloid-beta protein and AuNPs. The scanning spectra of SPR obtained from AuNPs of size 20 and 30nm were measured at wave length 526 nm and 533 nm respectively (Mirsadeghi et al., 2015). In the same context, adsorbing coronavirus proteins onto AuNPs resulted in changes in SPR bands and can easily detect the presence of virus (C et al., 2017; Mokhtarzadeh et al., 2017). Similarly, silver nanoparticles (AgNPs) showed the SPR behavior as AuNPs, in addition to electromagnetic properties, which was beneficial in the detection different viruses such hepatitis B virus by using Raman scattering detector (Li et al., 2013).

Magnetic NPs were examined for the ability of detection of viral infections. Nano $MnFe_2O_4$ was used in the detection of Influenza A virus by collecting the antibodies and extracting it using magnetic field, and the virus was visually detected using fluorescence microscope (LY et al., 2014). Another application of magnetic nanoparticles (MNPs) is in the immunoassay quantitative detection of influenza A virus subtype H1N1. Rapid interaction occurred between IgG Abs and MNPs resulted in surface excitation and attachment to H1N1 nucleoprotein, that could be further measured using spectroscopic biosensor (Wu et al., 2019) . Also, magnetic nanosorbent was used in immunoassay sandwich technique in which AuNPs acted as active surface for electrochemical detection of anti-Hepatitis B surface antigen (HBsAg) linked to IgG antibodies, as alternative method to standard enzyme linked immunosorbent assay (ELISA) (A et al., 2010).

2.2. Nano-biosensors

Recently, nanotechnological techniques have been utilized in re-designing diagnostic tools such as nanosensor-based techniques, which provide rapid and accurate tool in detecting low concentrations of viral load before the clear onset of symptoms (Fig. 2) (Abdeldayem et al., 2022). The analyte to be identified using nano-biosensor is either SARS-CoV-2 spike antigen, viral genome or SARS-CoV-2 antibodies. Several nanoparticle components as gold nanoparticle (AuNPs), quartz dot (Qdot), nanowire (NW) or graphene could be used in the detection. Biosensors are classified based on different transducing system technologies into electrical, optical and piezoelectric, and finally the signal is captured using detector.

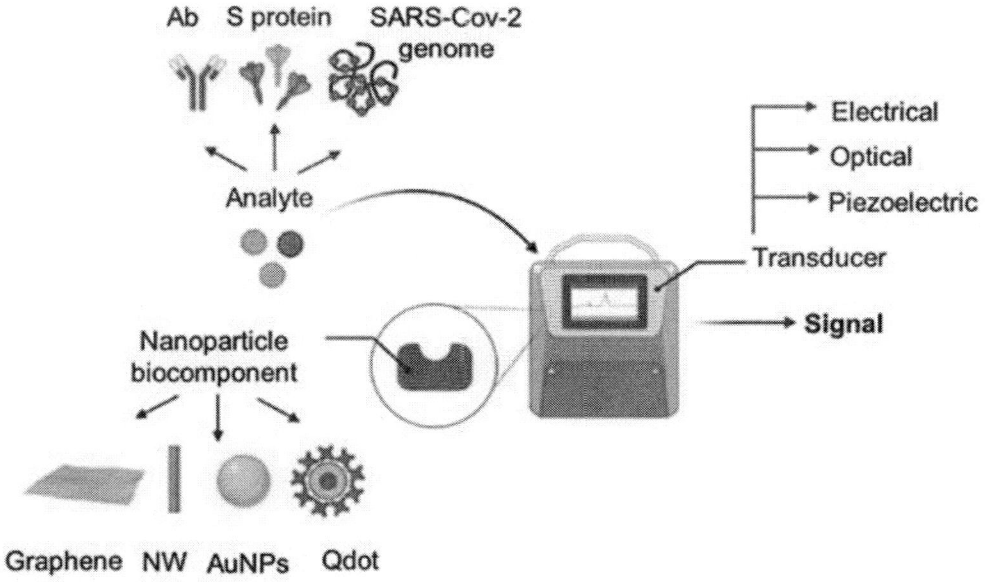

Figure 2. Nano-biosensor for the detection of COVID-19.

2.2.1. Electrochemical biosensors

These are considered the standard platform for the manufacturing process of biosensors (Thévenot et al., 2001), which are applicable in the examinations of biological targets such as nucleic acids, proteins, and cancer biomarkers (Kuralay et al., 2018; S et al., 2018). Electrochemical biosensors focus on monitoring any differences in the dielectric properties on the transducer surface, in addition to alteration in shape, dimension or charge distribution depending on amperometric, potentiometric, impedimetric and cyclic voltammetry fundamentals (Aydın, 2020; Chakraborty et al., 2019; Cordeiro et al., 2019). Several electrochemical detection mechanisms were applied for the detection of influenza and coronaviruses, including field effect transistor (FET), electrochemical impedance spectroscopy (EIS), bioelectric recognition assay (BERA) and conductometry (Chan et al., 2017; F et al., 2012; Hassen et al., 2011; L et al., 2014; LA & S, 2019; Mubarok et al., 2017; R et al., 2017; R. Wang et al., 2010; Wicklein et al., 2013). COVID-19 patient's saliva samples were analyzed for SARS-CoV-2 spike antigen using fast detector biosensor. The sensor was designed using fluorine-doped tin oxide (FTO) electrode in conjunction with AuNPs, whereas COVID-19 monoclonal antibody (nCOVID-19Ab) was immobilized and the alteration in the electric conductivity is measured. Various detection methods could be applied as cyclic voltammetry (CV) and differential pulse voltammetry (DPV) in identifying lower concentrations of COVID-19 up to 10 fM of nCOVID-19 Ag. The use of antibodies highlighted the importance of interaction with its corresponding antigen. Those biosensors efficiently detected COVID-19 with high accuracy, simplicity and at lower cost (Mahari et al., 2020).

2.2.2. Optical biosensors

Optical biosensors are multipurposed instruments cable of detecting multiple biological molecules using a single device (Garzón et al., 2019). The principle of action is through measuring the alteration in the optical properties on the transducer surface following the complex formation between the analyte and the recognition element. The analyte-molecule complex optical characteristics are measured either directly on the transducer surface, or indirectly using fluorophore or chromophore to identify the binding state and capture the signal (Soni et al., 2018). The most commonly used optical biosensors rely on the principles of surface plasmon resonance (SPR) and fluorescence, which are characterized by higher efficiency in detection if viruses such as SARS-CoV-2. The advantages of applying SPR biosensors are high sensitivity, accuracy and shorter detection time at nanoscale interface (Saylan, Erdem, et al., 2019; Shi et al., 2015). Moreover, localized surface plasmon resonance (LSPR) biosensing systems were beneficial in the diagnosis of several pathological and viral agents including coronaviruses and influenza viruses. LSPR biosensor was applied in the detection of targeted sequences of SARS-CoV-2 genome via nucleic acid hybridization (Qiu et al., 2020). Additionally, 3a gene which encodes a non-structural viral protein located on SARS-CoV RNA was detected through fluorescent biosensors (Waye et al., 2005). However, middle east respiratory syndrome coronavirus (MERS-CoV) diagnosis was performed using a label-free colorimetric assay relying on double-stranded DNA self-assembly shielded AuNPs under positive electrolyte (e.g., 0.1 M $MgCl_2$). Such colorimetric assay successfully detected up to 1 pmol/μL of 30 bp MERS-CoV (H. Kim et al., 2019).

2.2.3. Piezoelectric biosensors

Piezoelectric biosensors are widely used in medicinal applications including the detection of biological targets such as bacteria, cells, and hormones (Bunroddith et al., 2018; Yu et al., 2018).

Transducers utilized in these biosensors vibrate upon application of external electrical field. The immobilized biological target and the mass of the crystal cause variation in the resonance frequency when it contacts the analyte. Such alteration in the electrical signal and the mass difference could be recorded (Pohanka, 2018; Ricci et al., 2012; Saylan et al., 2017) Mechanical excitation of the probe results in resonance and alteration in the mass and viscoelasticity on the surface can be measured via tracing the frequency and modifying a quartz crystal resonator (Saylan, Akgönüllü, et al., 2019). Although such modification in the detection mechanism enabled label-free detection of molecules, it resulted in less precise measurements and added a level of complexity to the detection technique (Prasad, 2014). Piezoelectric immunosensors were previously used during SARS-CoV outbreak, and recently a modified nano-biosensor was fabricated to diagnose SARS-CoV-2 (Zuo et al., 2004).The combination of a quartz crystal microbalance (QCM)-based technique with piezoelectric biosensors was proved to be beneficial in the detection of various respiratory pathogens including influenza virus and SARS-CoV in oral swab samples (Jiang et al., 2015; Rahula Albano et al., 2018). QCM biosensor was designed for the detection of avian influenza using a 3-dimensional (3D) nano-well on a surface of gold (Au) (R. Wang & Li, 2013). Moreover, QCM-based nano-biosensor was used applied in the detection of SARS-CoV-2 spike protein collected using oral swab samples with high sensitivity reaches nanograms. The benefits of this methods are the rapid response and high sensitivity without the need for previous labeling (Pandey, 2020).

3. Prevention of COVID-19 infection

COVID-19 transmission occurs mostly as the result of person-to-person contact of the infected respiratory droplets. The healthcare systems recommend the use of facial masks, gloves, personal and protective equipment (PPE) kits which act as barrier against airborne droplets (Campos et al., 2020). Additional precautionary measures include insuring social distancing more than 1m, washing hands, using disinfectants, cleaning contamination zones, and isolating the infected patient (Greig et al., n.d.; Huang et al., 2020; J. Y. Kim et al., 2015). Nanomaterials have been efficiently utilized in the synthesis of some chemical and biological antimicrobial protective agents (Campos et al., 2020).

3.1. Nano-masks

Previous researches were conducted to explore the efficiency of different type of mask's fabrics in preventing COVID-19 transmission. It was reported that the single fabric mask with holes was not effective in filtering SARS-CoV-2 compared to doubled-fabric masks, prepared from silk and cotton, chiffon and cotton or flannel and cotton masks, which were more efficient with protective efficiency reached 90% in filtering SARS-CoV-2. Charged polyvinylidene difluoride (PVDF) nanofiber was proved to prevent the rapid transmission of coronavirus specially COVID-19 (Leung & Sun, 2020a, 2020b; Zhong et al., 2020).

Owing to the nanoparticle size of SARS-CoV-2, the viral-contaminated aerosol obtained from cough, sneeze or breath of the COVID-19 infected patients could be easily suspended in the air for long period of time. The national institute for occupational safety and health (NIOSH) has recommended the use of standard N95 and N99 masks in the highly contaminated environments, which can efficiently trap up to 300 nm aerosols, but ineffective against ultra-fine particles with size less than 300nm. Therefore, the development of air filters capable of capturing SARS-CoV-2 contaminated aerosol is essential for hindering the viral spread (Leung & Sun, 2020a). Nanoengineering techniques were applied in the manufacture of masks, gloves and PPE kits, producing antimicrobial, hydrophobic and comfortable fabric products. The combination of nanomaterials with textile fibers provides an additional antimicrobial activity to the textile. Nano-silver (AgNPs) impregnated fabrics have been utilized in the synthesis of face masks, lab coats, gowns and hospital curtains owing to its antimicrobial properties. Furthermore, NPs are characterized by long-term controlled release with provides regulated antimicrobial activity of the fabrics (Bhutta et al., 2020).

Moreover, graphene has been used in combination with copper and AgNPs to promote the antimicrobial properties of PPE fabrics. The integration of graphene within the fabric enhances its mechanical strength, flexibility, flame resistance besides the antimicrobial activity (Bhattacharjee et al., 2019). It was also reported that nanoengineered medical aprons and PPE kits are characterized by hydrophobicity, breathability, in addition to the inhibition of growth of microorganism on the textile surface. Nanoscale quaternary ammonium salts, peptides and polymers can also inhibit the oxidation of microbial membranes and thus modulate their growth (Rai et al., 2021). Modification in the manufacturing designs and techniques of facial masks includes coating with NPs which possess antiviral properties such as nano-silver or nano-copper. The nanocoated masks provide higher protection without influencing the breathability or the hydrophobicity of the fabric (Campos et al., 2020; Chaudhary et al., 2021). Furthermore, respiratory facial masks incorporated with AgNPs or coated with silica composite nanocoating were proved to have virucidal properties. Several research groups have modified face masks to provided proper protection against SARS-CoV-2 fatal effects. Fabrication of face masks using cellulose nanofibers produced breathable mask with disposable filter cartridge and allowed precise filtration of particles smaller than 100 nm. Moreover, the incorporation of microporous conductive graphene foam in the manufacture of facial masks induced a conduction of electric charges which traps and kills pathogenic microorganisms (Talebian et al., 2020) (Fig. 3). Nano-masks are synthesized using silver or copper nanoparticles which are impregnated within cellulose nanofiber. The mask is characterized by nanogaps that can trap particles > 100 nm, improved breathability, hydrophobicity and virucidal activity.

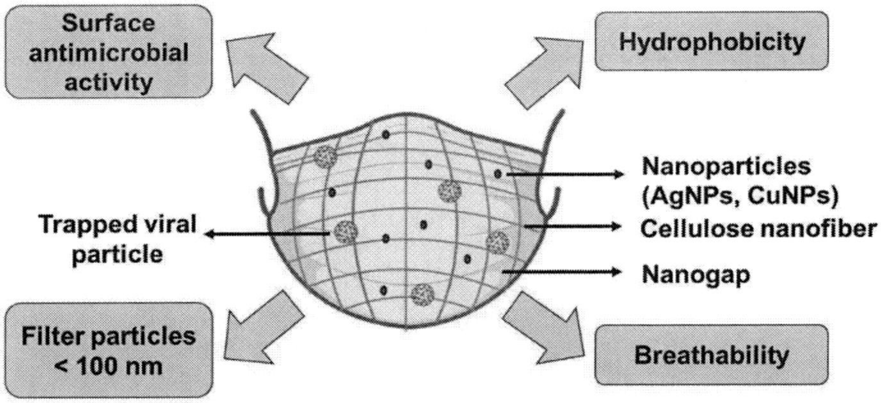

Figure 3. Nanoengineered masks with surface antimicrobial properties.

3.2. Sanitizers and Disinfectants

The use of sanitizers and disinfectants was recommended as one of the precautionary measures to restrain the transmission of microbes among surfaces. Nanotechnological innovations have provided significant support in the development of disinfectant agents. MNPs are known with their antiviral activity such as silver, copper, and titanium were presented as alternatives for the commonly used chemical disinfectants (Talebian et al., 2020). Nano-silver based multiuse sanitizer has been prepared using nano-colloidal technique. This environmental-friendly non-irritating sanitizer was proved to possess antiviral, antibacterial and antifungal properties. Another disinfectant was synthesized using silver ions and titanium dioxide to sterilize surfaces contaminated with coronavirus. Additionally, the antimicrobial nano-polymer based disinfectants offer several advantages over chemical disinfectants such as biodegradability and absence of inflammability (Chaudhary et al., 2021; Rai et al., 2021).

4. Developing vaccination

The common traditional vaccines which depend on live-attenuated or inactivated pathogens have several harmful side effects from reflection to pathogenic virulence or guide to a weak immune response, so the non-traditional vaccines based on NPs can enhance strong immunogenicity and antigen presentation by controlling the surface, shape, size and functional groups (Al-Halifa et al., 2019). Nanotechnology can be an ideal solution to overcome the drawbacks of inactivated vaccines (Alphandéry, 2020). The generated NPs from virus-like, polymers (biocompatible and biodegradable) and liposomes have special physicochemical properties increase its ability to use as vaccines delivery platform (Dhakal & Renukaradhya, 2019).

Nanotechnology applications revealed new area for the use of multifunctional agents with programmable characteristics, which can create new treatment strategies against viral infections as COVID-19 (Song et al., 2019). The rationale after using nanotechnology in the production of vaccines and other therapeutic approaches is that the nano-sized particles could easily enter into the cells resulting in stimulation of antigen expression against the delivered nucleic acid (as in DNA (deoxyribonucleic acid) and mRNA (messenger ribonucleic acid) vaccines) or in direct targeting to the immune cells, enabling the delivery of antigens (as in subunit vaccines). Another advantage is that NPs have shielded properties that protect encapsulated materials or anti-infectious agents from degradation (Chung et al., 2020). NPs can be engineered to inhibit viral entry into the host cells via blocking the interaction between the virus and cell's receptors or to directly inactivate the viral particles (Nasrollahzadeh et al., 2019).

Several approaches relied on encapsulation of the genomic material or peptide antigen within NPs including lipid nanoparticle (LNP) or viral vectors as adenovirus (Ads) were applied. BioNTech/Pfizer and Moderna vaccine is based on mRNA encapsulation within LNPs, however, Oxford/Astrazeneca and CanSino vaccines relied on incorporation of antigen-encoding sequences located within the DNA and transferred by Ads. While Novavax applied recombinant nanoparticle technology using S protein of SARS-CoV-2 to produce virus like particle (VLP) nanoparticle-based vaccine (Folegatti et al., 2020; Jackson et al., 2020; Zhu et al., 2020).

Besides antigen delivery into the cell, NPs could also assist codelivery of adjuvants to activate the immune response. Adjuvants are usually administered with the vaccine to enhance the immune response through activation of molecular receptors that recognize external invaders and pathogens.

Encapsulation of the antigen and the adjuvant within the same nanoparticle authorizes synchronized targeted delivery of the nano-capsule to the same antigen presenting cell (APC) (Coffman et al., 2010; Kreutz et al., 2012; McKee et al., 2010). Absence of temporal synchronization, where the antigen and the adjuvant are picked by the APCs at separate times, could result in autoimmune responses against host's proteins, as the adjuvant could stimulate APCs that are prepared against the self-antigens and not the delivered anonymous antigen. Various techniques have been followed to ensure the codelivery of SARS-CoV-2 antigens and adjuvants to the cells including: first, antigen and adjuvant encapsulation or incorporation within the same nanoparticle, second, direct antigen-adjuvant binding, and third, via dual functionality of the adjuvant as immunostimulatory molecule and delivery vehicle. Consequently, nanotechnology permits the targeted delivery of vaccines, thus improve their efficiencies (Pati et al., 2018; Z. B. Wang & Xu, 2020).

Conclusion

Rapid transmission of COVID-19 across the globe became a disturbing issue, which can be prevented by the early and accurate detection of the infected patients. Nanotechnology has developed new approaches for diagnosis, prevention of the infection and synthesis of vaccines against COVID-19. NPs are considered perfect candidates for the rapid detection of SARS-CoV-2 relying on their nanoscale size, which enhances the direct interaction with the viral particles. Moreover, NPs have great influence on the effectiveness and sensitivity of biosensing applications. The fabrication of nano-biosensors facilitates the accurate and rapid detection of low viral concentrations in few seconds, with high degree of sensitivity essential for mass screening during the outbreaks. The utilization of nanoengineering approaches in the synthesis of facial masks and personal protective equipment inhibits the growth of microorganisms on the textile surface, resulting in fabrics with better hydrophobicity, breathability and antimicrobial activities. In nano-vaccines preparation, the generated NPs prepared from liposome or viral vectors enhance viral targeting, thus promotes vaccine delivery, decreases antigen's degradation and boosts the immunogenicity. Consequently, nanotechnology could be considered an ideal solution to fight against COVID-19 pandemic.

References

A, de la E.-M., M, M. C., C, S.-E., B, D.-F., J, F.-S., Á, G.-F., & A, M. (2010). Gold nanoparticle-based electrochemical magnetoimmunosensor for rapid detection of anti-hepatitis B virus antibodies in human serum. *Biosensors & Bioelectronics*, *26*(4), 1710–1714. https://doi.org/10.1016/J.BIOS.2010.07.069

Abdeldayem, O. M., Dabbish, A. M., Habashy, M. M., Mostafa, M. K., Elhefnawy, M., Amin, L., Al-Sakkari, E. G., Ragab, A., & Rene, E. R. (2022). Viral outbreaks detection and surveillance using wastewater-based epidemiology, viral air sampling, and machine learning techniques: A comprehensive review and outlook. *The Science of the Total Environment*, *803*, 149834. https://doi.org/10.1016/J.SCITOTENV.2021.149834

Al-Halifa, S., Gauthier, L., Arpin, D., Bourgault, S., & Archambault, D. (2019). Nanoparticle-Based Vaccines Against Respiratory Viruses. *Frontiers in Immunology*, *10*(JAN). https://doi.org/10.3389/FIMMU.2019.00022

Alphandéry, E. (2020). The Potential of Various Nanotechnologies for Coronavirus Diagnosis/Treatment Highlighted through a Literature Analysis. *Bioconjugate Chemistry*, *31*(8), 1873–1882. https://doi.org/10.1021/ACS.BIOCONJCHEM.0C00287

Aydın, E. B. (2020). Highly sensitive impedimetric immunosensor for determination of interleukin 6 as a cancer biomarker by using conjugated polymer containing epoxy side groups modified disposable ITO electrode. *Talanta*, *215*, 120909. https://doi.org/10.1016/J.TALANTA.2020.120909

Bhattacharjee, S., Joshi, R., Chughtai, A. A., & Macintyre, C. R. (2019). Graphene Modified Multifunctional Personal Protective Clothing. *Advanced Materials Interfaces*, *6*(21). https://doi.org/10.1002/ADMI.201900622

Bhutta, Z. A., Kanwal, A., & Ali, M. (2020). Nanotechnology based approaches for combatting COVID-19 viral infection. *Nano Express*, *1*(2), 022003. https://doi.org/10.1088/2632-959X/ABB714

Bunroddith, K., Viseshakul, N., Chansiri, K., & Lieberzeit, P. (2018). QCM-based rapid detection of PCR amplification products of Ehrlichia canis. *Analytica Chimica Acta*, *1001*, 106–111. https://doi.org/10.1016/J.ACA.2017.10.037

C, L., P, W., MA, G., AA, W., & P, Z. (2017). Plasmonics-Based Detection of Virus Using Sialic Acid Functionalized Gold Nanoparticles. *Methods in Molecular Biology (Clifton, N.J.)*, *1571*, 109–116. https://doi.org/10.1007/978-1-4939-6848-0_7

Campos, E. V. R., Pereira, A. E. S., De Oliveira, J. L., Carvalho, L. B., Guilger-Casagrande, M., De Lima, R., & Fraceto, L. F. (2020). How can nanotechnology help to combat COVID-19? Opportunities and urgent need. *Journal of Nanobiotechnology*, *18*(1). https://doi.org/10.1186/S12951-020-00685-4

Cascella, M., Rajnik, M., Aleem, A., Dulebohn, S. C., & Napoli, R. Di. (2021). Features, Evaluation, and Treatment of Coronavirus (COVID-19). *StatPearls*.

Chakraborty, A., Tibarewala, D. N., & Barui, A. (2019). Impedance-based biosensors. *Bioelectronics and Medical Devices: From Materials to Devices - Fabrication, Applications and Reliability*, 97–122. https://doi.org/10.1016/B978-0-08-102420-1.00005-4

Chan, C., Shi, J., Fan, Y., & Yang, M. (2017). A microfluidic flow-through chip integrated with reduced graphene oxide transistor for influenza virus gene detection. *Sensors and Actuators B: Chemical*, *251*, 927–933. https://doi.org/10.1016/J.SNB.2017.05.147

Chaudhary, V., Royal, A., Chavali, M., & Yadav, S. K. (2021). Advancements in research and development to combat COVID-19 using nanotechnology. *Nanotechnology for Environmental Engineering*, *6*(1), 1–15. https://doi.org/10.1007/S41204-021-00102-7/TABLES/3

Chung, Y. H., Beiss, V., Fiering, S. N., & Steinmetz, N. F. (2020). Covid-19 vaccine frontrunners and their nanotechnology design. *ACS Nano*, *14*(10), 12522–12537. https://doi.org/10.1021/ACSNANO.0C07197/SUPPL_FILE/NN0C07197_SI_001.PDF

Coffman, R. L., Sher, A., & Seder, R. A. (2010). Vaccine adjuvants: putting innate immunity to work. *Immunity*, *33*(4), 492–503. https://doi.org/10.1016/J.IMMUNI.2010.10.002

Cordeiro, T. A. R., Gonçalves, M. V. C., Franco, D. L., Reis, A. B., Martins, H. R., & Ferreira, L. F. (2019). Label-free electrochemical impedance immunosensor based on modified screen-printed gold electrodes for the diagnosis of canine visceral leishmaniasis. *Talanta*, *195*, 327–332. https://doi.org/10.1016/J.TALANTA.2018.11.087

Dabbish, A. M., Yonis, N., Salama, M., Essa, M. M., & Qoronfleh, M. W. (2021). Inflammatory pathways and potential therapies for COVID-19: A mini review. *European Journal of Inflammation*, *19*. https://doi.org/10.1177/20587392211002986

Dhakal, S., & Renukaradhya, G. J. (2019). Nanoparticle-based vaccine development and evaluation against viral infections in pigs. *Veterinary Research 2019 50:1*, *50*(1), 1–14. https://doi.org/10.1186/S13567-019-0712-5

Draz, M. S., & Shafiee, H. (2018). Applications of gold nanoparticles in virus detection. *Theranostics*, *8*(7), 1985. https://doi.org/10.7150/THNO.23856

F, S., J, W., Z, X., Y, W., Q, C., X, L., X, J., L, L., M, Y., X, G., & T, Z. (2012). Rapid flu diagnosis using silicon nanowire sensor. *Nano Letters*, *12*(7), 3722–3730. https://doi.org/10.1021/NL301516Z

Folegatti, P. M., Ewer, K. J., Aley, P. K., Angus, B., Becker, S., Belij-Rammerstorfer, S., Bellamy, D., Bibi, S., Bittaye, M., Clutterbuck, E. A., Dold, C., Faust, S. N., Finn, A., Flaxman, A. L., Hallis, B., Heath, P., Jenkin, D., Lazarus, R., Makinson, R., … Yau, Y. (2020). Safety and immunogenicity of the ChAdOx1 nCoV-19 vaccine against SARS-CoV-2: a preliminary report of a phase 1/2, single-blind, randomised controlled trial. *Lancet (London, England)*, *396*(10249), 467. https://doi.org/10.1016/S0140-6736(20)31604-4

Garzón, V., Pinacho, D. G., Bustos, R. H., Garzón, G., & Bustamante, S. (2019). Optical Biosensors for Therapeutic Drug Monitoring. *Biosensors 2019, Vol. 9, Page 132*, *9*(4), 132. https://doi.org/10.3390/BIOS9040132

Greig, C., Levine, B., Long, G., Mekjavic, I., Montgomery, H., Morgan, P., Roiz De Sa, D., Wilkes, M., Wilson, J., Society, T. P., Mackenzie, A., Addison, T., & Mckay, L. (n.d.). *Returning to human testing: lab and field*.

H, H., C, F., M, L., HL, N., FB, W., H, W., R, W., J, X., X, Z., X, Z., & J, H. (2020). COVID-19: A Call for Physical Scientists and Engineers. *ACS Nano*, *14*(4), 3747–3754. https://doi.org/10.1021/ACSNANO.0C02618

Hamdy, A., Mostafa, M. K., & Nasr, M. (2019). Techno-economic estimation of electroplating wastewater treatment using zero-valent iron nanoparticles: batch optimization, continuous feed, and scaling up studies. *Environmental Science and Pollution Research International*, *26*(24), 25372–25385. https://doi.org/10.1007/S11356-019-05850-3

Hassen, W. M., Duplan, V., Frost, E., & Dubowski, J. J. (2011). Quantitation of influenza A virus in the presence of extraneous protein using electrochemical impedance spectroscopy. *Electrochimica Acta*, *56*(24), 8325–8328. https://doi.org/10.1016/J.ELECTACTA.2011.07.009

Huang, H., Fan, C., Li, M., Nie, H. L., Wang, F. B., Wang, H., Wang, R., Xia, J., Zheng, X., Zuo, X., & Huang, J. (2020). COVID-19: A Call for Physical Scientists and Engineers. *ACS Nano*, *14*(4), 3747–3754. https://doi.org/10.1021/ACSNANO.0C02618

Jackson, L. A., Anderson, E. J., Rouphael, N. G., Roberts, P. C., Makhene, M., Coler, R. N., McCullough, M. P., Chappell, J. D., Denison, M. R., Stevens, L. J., Pruijssers, A. J., McDermott, A., Flach, B., Doria-Rose, N. A., Corbett, K. S., Morabito, K. M., O'Dell, S., Schmidt, S. D., Swanson, P. A., … Beigel, J. H. (2020). An mRNA Vaccine against SARS-CoV-2 - Preliminary Report. *The New England Journal of Medicine*, *383*(20), 1920–1931. https://doi.org/10.1056/NEJMOA2022483

Jiang, Y., Tan, C. Y., Tan, S. Y., Wong, M. S. F., Chen, Y. F., Zhang, L., Yao, K., Gan, S. K. E., Verma, C., & Tan, Y. J. (2015). SAW sensor for Influenza A virus detection enabled with efficient surface functionalization. *Sensors and Actuators B: Chemical*, *209*, 78–84. https://doi.org/10.1016/J.SNB.2014.11.103

Kim, H., Park, M., Hwang, J., Kim, J. H., Chung, D. R., Lee, K. sung, & Kang, M. (2019). Development of Label-Free Colorimetric Assay for MERS-CoV Using Gold Nanoparticles. *ACS Sensors*, *4*(5), 1306–1312. https://doi.org/10.1021/ACSSENSORS.9B00175/SUPPL_FILE/SE9B00175_SI_001.PDF

Kim, J. Y., Song, J. Y., Yoon, Y. K., Choi, S. H., Song, Y. G., Kim, S. R., Son, H. J., Jeong, S. Y., Choi, J. H., Kim, K. M., Yoon, H. J., Choi, J. Y., Kim, T. H., Choi, Y. H., Kim, H. Bin, Yoon, J. H., Lee, J., Eom, J. S., Lee, S. O., … Cheong, H. J. (2015). Middle East Respiratory Syndrome Infection Control and Prevention Guideline for Healthcare Facilities. *Infection & Chemotherapy*, *47*(4), 278. https://doi.org/10.3947/IC.2015.47.4.278

KK, J. (2005). Nanotechnology in clinical laboratory diagnostics. *Clinica Chimica Acta; International Journal of Clinical Chemistry*, *358*(1–2), 37–54. https://doi.org/10.1016/J.CCCN.2005.03.014

Kreutz, M., Giquel, B., Hu, Q., Abuknesha, R., Uematsu, S., Akira, S., Nestle, F. O., & Diebold, S. S. (2012). Antibody-Antigen-Adjuvant Conjugates Enable Co-Delivery of Antigen and Adjuvant to Dendritic Cells in Cis but Only Have Partial Targeting Specificity. *PLoS ONE*, *7*(7). https://doi.org/10.1371/JOURNAL.PONE.0040208

Kuralay, F., Dükar, N., & Bayramlı, Y. (2018). Poly-L-lysine Coated Surfaces for Ultrasensitive Nucleic Acid Detection. *Electroanalysis*, *30*(7), 1556–1565. https://doi.org/10.1002/ELAN.201800153

L, K., L, N., MA, R., M, Z., M, M., D, H., O, Z., P, K., V, A., & R, K. (2014). 3D printed chip for electrochemical detection of influenza virus labeled with CdS quantum dots. *Biosensors & Bioelectronics*, *54*, 421–427. https://doi.org/10.1016/J.BIOS.2013.10.031

LA, L., & S, E. (2019). An electrochemical immunosensor for the corona virus associated with the Middle East respiratory syndrome using an array of gold nanoparticle-modified carbon electrodes. *Mikrochimica Acta*, *186*(4). https://doi.org/10.1007/S00604-019-3345-5

Leung, W. W. F., & Sun, Q. (2020a). Charged PVDF multilayer nanofiber filter in filtering simulated airborne novel coronavirus (COVID-19) using ambient nano-aerosols. *Separation and Purification Technology*, *245*, 116887. https://doi.org/10.1016/J.SEPPUR.2020.116887

Leung, W. W. F., & Sun, Q. (2020b). Electrostatic charged nanofiber filter for filtering airborne novel coronavirus (COVID-19) and nano-aerosols. *Separation and Purification Technology*, *250*, 116886. https://doi.org/10.1016/J.SEPPUR.2020.116886

Li, M., Cushing, S. K., Liang, H., Suri, S., Ma, D., & Wu, N. (2013). Plasmonic Nanorice Antenna on Triangle Nanoarray for Surface-Enhanced Raman Scattering Detection of Hepatitis B Virus DNA. *Analytical Chemistry*, *85*(4), 2072–2078. https://doi.org/10.1021/AC303387A

LY, H., JC, C., YC, T., CC, H., CP, C., CS, Y., & GB, L. (2014). Magnetic nanoparticle-based immunoassay for rapid detection of influenza infections by using an integrated microfluidic system. *Nanomedicine : Nanotechnology, Biology, and Medicine*, *10*(4), 819–829. https://doi.org/10.1016/J.NANO.2013.11.009

Mahari, S., Roberts, A., Shahdeo, D., & Gandhi, S. (2020). eCovSens-Ultrasensitive Novel In-House Built Printed Circuit Board Based Electrochemical Device for Rapid Detection of nCovid-19 antigen, a spike protein domain 1 of SARS-CoV-2. *BioRxiv*, 2020.04.24.059204. https://doi.org/10.1101/2020.04.24.059204

Mahmoud, A. S., Mostafa, M. K., & Peters, R. W. (2021). A prototype of textile wastewater treatment using coagulation and adsorption by Fe/Cu nanoparticles: Techno-economic and scaling-up studies: *Https://Doi.Org/10.1177/18479804211041181*, *11*, 1–21. https://doi.org/10.1177/18479804211041181

McKee, A. S., MacLeod, M. K., Kappler, J. W., & Marrack, P. (2010). Immune mechanisms of protection: can adjuvants rise to the challenge? *BMC Biology*, *8*(1), 37. https://doi.org/10.1186/1741-7007-8-37

Mirsadeghi, S., Dinarvand, R., Ghahremani, M. H., Hormozi-Nezhad, M. R., Mahmoudi, Z., Hajipour, M. J., Atyabi, F., Ghavami, M., & Mahmoudi, M. (2015). Protein corona composition of gold nanoparticles/nanorods affects amyloid beta fibrillation process. *Nanoscale*, *7*(11), 5004–5013. https://doi.org/10.1039/C4NR06009A

Mokhtarzadeh, A., Eivazzadeh-Keihan, R., Pashazadeh, P., Hejazi, M., Gharaatifar, N., Hasanzadeh, M., Baradaran, B., & de la Guardia, M. (2017). Nanomaterial-based biosensors for detection of pathogenic virus. *TrAC Trends in Analytical Chemistry*, *97*, 445–457. https://doi.org/10.1016/J.TRAC.2017.10.005

Mostafa, M. K., & Peters, R. W. (2016). Improve effluent water quality at Abu-Rawash wastewater treatment plant with the application of coagulants. *Water and Environment Journal*, *30*(1–2), 88–95. https://doi.org/10.1111/WEJ.12161

Mubarok, A. Z., Mani, V., Huang, C. H., Chang, P. C., & Huang, S. T. (2017). Label-free electrochemical detection of neuraminidase activity: A facile whole blood diagnostic probe for infectious diseases. *Sensors and Actuators B: Chemical*, *252*, 641–648. https://doi.org/10.1016/J.SNB.2017.06.061

N, R., M, B., A, G., H, Z., S, A., Y, F., R, D., M, R., S, R., M, S., & RS, V. (2020). Point-of-Use Rapid Detection of SARS-CoV-2: Nanotechnology-Enabled Solutions for the COVID-19

Pandemic. *International Journal of Molecular Sciences*, *21*(14), 1–23. https://doi.org/10.3390/IJMS21145126

Nasrollahzadeh, M., Sajadi, S. M., Sajjadi, M., & Issaabadi, Z. (2019). An Introduction to Nanotechnology. *Interface Science and Technology*, *28*, 1–27. https://doi.org/10.1016/B978-0-12-813586-0.00001-8

Pandey, L. M. (2020). Design of engineered surfaces for prospective detection of SARS-CoV-2 using quartz crystal microbalance-based techniques. *Expert Review of Proteomics*, *17*(6). https://doi.org/10.1080/14789450.2020.1794831

Pati, R., Shevtsov, M., & Sonawane, A. (2018). Nanoparticle Vaccines Against Infectious Diseases. *Frontiers in Immunology*, *9*(OCT), 2224. https://doi.org/10.3389/FIMMU.2018.02224

Pohanka, M. (2018). Piezoelectric biosensor for the determination of Tumor Necrosis Factor Alpha. *Talanta*, *178*, 970–973. https://doi.org/10.1016/J.TALANTA.2017.10.031

Prasad, S. (2014). Nanobiosensors: the future for diagnosis of disease? *Nanobiosensors in Disease Diagnosis*, *3*, 1–10. https://doi.org/10.2147/NDD.S39421

Qiu, G., Gai, Z., Tao, Y., Schmitt, J., Kullak-Ublick, G. A., & Wang, J. (2020). Dual-Functional Plasmonic Photothermal Biosensors for Highly Accurate Severe Acute Respiratory Syndrome Coronavirus 2 Detection. *ACS Nano*, *14*(5), 5268–5277. https://doi.org/10.1021/ACSNANO.0C02439

R, S., S, H., & J, J. (2017). Label-free Detection of Influenza Viruses using a Reduced Graphene Oxide-based Electrochemical Immunosensor Integrated with a Microfluidic Platform. *Scientific Reports*, *7*. https://doi.org/10.1038/SREP42771

Rahula Albano, D. B., Shum, K., Tanner, J. A., & Sing Fung, Y. (2018). BS5.3 - Piezoelectric quartz crystal aptamer biosensor for detection and quantification of SARS CoV helicase protein. *Proceedings IMCS 2018*, 211–213. https://doi.org/10.5162/IMCS2018/BS5.3

Rai, M., Bonde, S., Yadav, A., Bhowmik, A., Rathod, S., Ingle, P., & Gade, A. (2021). Nanotechnology as a Shield against COVID-19: Current Advancement and Limitations. *Viruses*, *13*(7). https://doi.org/10.3390/V13071224

Ramakrishnan, S. G., Robert, B., Salim, A., Ananthan, P., Sivaramakrishnan, M., Subramaniam, S., Natesan, S., Suresh, R., Rajeshkumar, G., Maran, J. P., Al-Dhabi, N. A., Karuppiah, P., & Arasu, M. V. (2021). Nanotechnology based solutions to combat zoonotic viruses with special attention to SARS, MERS, and COVID 19: Detection, protection and medication. *Microbial Pathogenesis*, *159*, 105133. https://doi.org/10.1016/J.MICPATH.2021.105133

Rangayasami, A., Kannan, K., Murugesan, S., Radhika, D., Sadasivuni, K. K., Reddy, K. R., & Raghu, A. V. (2021). Influence of nanotechnology to combat against COVID-19 for global health emergency: A review. *Sensors International*, *2*, 100079. https://doi.org/10.1016/J.SINTL.2020.100079

Ricci, F., Adornetto, G., & Palleschi, G. (2012). A review of experimental aspects of electrochemical immunosensors. *Electrochimica Acta*, *84*, 74–83.

https://doi.org/10.1016/J.ELECTACTA.2012.06.033

Ruiz-Hitzky, E., Darder, M., Wicklein, B., Ruiz-Garcia, C., Martín-Sampedro, R., Real, G. del, & Aranda, P. (2020). Nanotechnology Responses to COVID-19. *Advanced Healthcare Materials*, *9*(19), 2000979. https://doi.org/10.1002/ADHM.202000979

S, R., GK, M., J, A., RK, M., KY, G., MAH, N., JL, M., & A, H. (2018). Carboxylic group riched graphene oxide based disposable electrochemical immunosensor for cancer biomarker detection. *Analytical Biochemistry*, *545*, 13–19. https://doi.org/10.1016/J.AB.2018.01.007

Saylan, Y., Akgönüllü, S., Yavuz, H., Ünal, S., & Denizli, A. (2019). Molecularly Imprinted Polymer Based Sensors for Medical Applications. *Sensors (Basel, Switzerland)*, *19*(6). https://doi.org/10.3390/S19061279

Saylan, Y., Erdem, Ö., Ünal, S., & Denizli, A. (2019). An Alternative Medical Diagnosis Method: Biosensors for Virus Detection. *Biosensors*, *9*(2). https://doi.org/10.3390/BIOS9020065

Saylan, Y., Yilmaz, F., Özgür, E., Derazshamshir, A., Yavuz, H., & Denizli, A. (2017). Molecular Imprinting of Macromolecules for Sensor Applications. *Sensors 2017, Vol. 17, Page 898*, *17*(4), 898. https://doi.org/10.3390/S17040898

Shi, L., Sun, Q., He, J., Xu, H., Liu, C., Zhao, C., Xu, Y., Wu, C., Xiang, J., Gu, D., Long, J., & Lan, H. (2015). Development of SPR biosensor for simultaneous detection of multiplex respiratory viruses. *Bio-Medical Materials and Engineering*, *26 Suppl 1*, S2207–S2216. https://doi.org/10.3233/BME-151526

Somvanshi, S. B., Kharat, P. B., Saraf, T. S., Somwanshi, S. B., Shejul, S. B., & Jadhav, K. M. (2020). Multifunctional nano-magnetic particles assisted viral RNA-extraction protocol for potential detection of COVID-19. *Materials Research Innovations*, 1. https://doi.org/10.1080/14328917.2020.1769350

Song, Z., Xu, Y., Bao, L., Zhang, L., Yu, P., Qu, Y., Zhu, H., Zhao, W., Han, Y., & Qin, C. (2019). From SARS to MERS, Thrusting Coronaviruses into the Spotlight. *Viruses*, *11*(1). https://doi.org/10.3390/V11010059

Soni, A., Surana, R. K., & Jha, S. K. (2018). Smartphone based optical biosensor for the detection of urea in saliva. *Sensors and Actuators B: Chemical*, *269*, 346–353. https://doi.org/10.1016/J.SNB.2018.04.108

Talebian, S., Wallace, G. G., Schroeder, A., Stellacci, F., & Conde, J. (2020). Nanotechnology-based disinfectants and sensors for SARS-CoV-2. *Nature Nanotechnology*, *15*(8), 618–621. https://doi.org/10.1038/S41565-020-0751-0

Tang, Z., Zhang, X., Shu, Y., Guo, M., Zhang, H., & Tao, W. (2021). Insights from nanotechnology in COVID-19 treatment. *Nano Today*, *36*, 101019. https://doi.org/10.1016/J.NANTOD.2020.101019

Thévenot, D. R., Toth, K., Durst, R. A., & Wilson, G. S. (2001). Electrochemical biosensors: recommended definitions and classification. *Biosensors and Bioelectronics*, *16*(1–2), 121–

131. https://doi.org/10.1016/S0956-5663(01)00115-4

Tu, H., Tu, S., Gao, S., Shao, A., & Sheng, J. (2020). Current epidemiological and clinical features of COVID-19; a global perspective from China. *The Journal of Infection*, *81*(1), 1. https://doi.org/10.1016/J.JINF.2020.04.011

W, G., C, W., J, C., & Y, X. (2013). Nanotechnology in the targeted drug delivery for bone diseases and bone regeneration. *International Journal of Nanomedicine*, *8*, 2305–2317. https://doi.org/10.2147/IJN.S44393

Wang, R., & Li, Y. (2013). Hydrogel based QCM aptasensor for detection of avian influenza virus. *Biosensors & Bioelectronics*, *42*(1), 148–155. https://doi.org/10.1016/J.BIOS.2012.10.038

Wang, R., Li, Y., Mao, X., Huang, T., & Lu, H. (2010). Magnetic bio-nanobeads and nanoelectrode based impedance biosensor for detection of avian influenza virus. *2010 IEEE International Conference on Nano/Molecular Medicine and Engineering, IEEE NANOMED 2010*, 214–217. https://doi.org/10.1109/NANOMED.2010.5749837

Wang, Z. B., & Xu, J. (2020). Better Adjuvants for Better Vaccines: Progress in Adjuvant Delivery Systems, Modifications, and Adjuvant–Antigen Codelivery. *Vaccines*, *8*(1). https://doi.org/10.3390/VACCINES8010128

Waye, M. M. Y., Law, P. T. W., Wong, C. H., Au, T. C. C., Chuck, C. P., Kong, S. K., Chan, P. K. S., To, K. F., Lo, A. W. I., Chan, J. Y. W., Suen, Y. K., Chan, H. Y. E., Fung, K. P., Sung, J. J. Y., Lo, Y. M. D., & Tsui, S. K. W. (2005). The 3a Protein of SARS-coronavirus Induces Apoptosis in Vero E6 Cells. *Conference Proceedings : ... Annual International Conference of the IEEE Engineering in Medicine and Biology Society. IEEE Engineering in Medicine and Biology Society. Annual Conference*, *2005*, 7482–7485. https://doi.org/10.1109/IEMBS.2005.1616242

Wicklein, B., Burgo, M. Á. M. del, Yuste, M., Carregal-Romero, E., Llobera, A., Darder, M., Aranda, P., Ortín, J., Real, G. del, Fernández-Sánchez, C., & Ruiz-Hitzky, E. (2013). Biomimetic Architectures for the Impedimetric Discrimination of Influenza Virus Phenotypes. *Advanced Functional Materials*, *23*(2), 254–262. https://doi.org/10.1002/ADFM.201200377

Wu, K., Liu, J., Saha, R., Su, D., Krishna, V. D., Cheeran, M., & Wang, J.-P. (2019). Detection of Influenza A Virus Nucleoprotein Through the Self-Assembly of Nanoparticles in Magnetic Particle Spectroscopy-Based Bioassays: A Method for Rapid, Sensitive, and Wash-free Magnetic Immunoassays. *Undefined*.

Yu, X., Chen, F., Wang, R., & Li, Y. (2018). Whole-bacterium SELEX of DNA aptamers for rapid detection of E.coli O157:H7 using a QCM sensor. *Journal of Biotechnology*, *266*, 39–49. https://doi.org/10.1016/J.JBIOTEC.2017.12.011

Zeroual, A., Harrou, F., Dairi, A., & Sun, Y. (2020). Deep learning methods for forecasting COVID-19 time-Series data: A Comparative study. *Chaos, Solitons & Fractals*, *140*, 110121. https://doi.org/10.1016/J.CHAOS.2020.110121

Zhong, H., Zhu, Z., Lin, J., Cheung, C. F., Lu, V. L., Yan, F., Chan, C. Y., & Li, G. (2020).

Reusable and Recyclable Graphene Masks with Outstanding Superhydrophobic and Photothermal Performances. *ACS Nano*, *14*(5), 6213–6221. https://doi.org/10.1021/ACSNANO.0C02250

Zhu, F. C., Guan, X. H., Li, Y. H., Huang, J. Y., Jiang, T., Hou, L. H., Li, J. X., Yang, B. F., Wang, L., Wang, W. J., Wu, S. P., Wang, Z., Wu, X. H., Xu, J. J., Zhang, Z., Jia, S. Y., Wang, B. Sen, Hu, Y., Liu, J. J., … Chen, W. (2020). Immunogenicity and safety of a recombinant adenovirus type-5-vectored COVID-19 vaccine in healthy adults aged 18 years or older: a randomised, double-blind, placebo-controlled, phase 2 trial. *Lancet (London, England)*, *396*(10249), 479. https://doi.org/10.1016/S0140-6736(20)31605-6

Zuo, B., Li, S., Guo, Z., Zhang, J., & Chen, C. (2004). Piezoelectric immunosensor for SARS-associated coronavirus in sputum. *Analytical Chemistry*, *76*(13), 3536–3540. https://doi.org/10.1021/AC035367B

634816 Invited Talk: Taking Electrochemical Biomolecular Sensors from the Bench to the Clinic

Abstract:

To put disease-related biomarkers to work in the clinic, new high-performance technologies are needed to enable rapid and sensitive analysis of clinical specimens. Electrochemical methods providing low cost and direct biomarker readout have attracted a great deal of attention for this application. In our laboratories, we have used controlled nanostructuring of electrode surfaces to enhance biomolecular capture rates and efficiencies (*J. Am. Chem. Soc.* 2017). By immobilizing biomolecules on nanostructured surfaces, high-performance biomolecular detection systems can be developed to facilitate rapid biomarker analysis and the detection of infectious pathogens (*J. Phys. Chem. C.* 2016; *Nano Lett.* 2017). We have developed electrochemical assays that are able to detect nucleic acids, proteins and small molecules, with universally high sensitivity levels and have applied this approach in a variety of areas including noninvasive monitoring via Liquid Biopsy (*Nature Chemistry* 2015) and antimicrobial resistance assessment (*Nature Chemistry* 2020). Recently we have developed reagentless sensors that are powerful detectors for viral sensing (*J. Am. Chem. Soc.* 2021) and in vivo sensing (*Nature Chemistry*, 2021). This talk will summarize the development of these sensors, their application to a variety of clinical problems, and their evolution into a clinical-grade products.

627855 Sensitive Electrochemical Detection of Pathogenic *E. coli*

Abstract:

Pathogenic *E. coli* are a significant threat as a foodborne pathogen and a cause of urinary tract infections. Conventional methods to detect these microbes require centralized facilities and trained personnel, as well as hours to days to complete. We have developed a sensitive electrochemical platform to covalently capture, detect, and quantify viable pathogenic *E. coli* from complex matrices with a detection limit of 12 CFU/mL and a linear range of detection up to 10^7 CFU/mL. The technology enables quantification within two hours on disposable electrodes, even from milk and artificial urine. Endogenous *E. coli* were successfully detected from commonly-contaminated samples, including eggs, raw chicken, spinach, romaine lettuce, and an infected urine sample. The *E. coli* quantification was as sensitive as the current gold-standard of colony counting, but with significantly less error. This platform represents a major step in the development of field-deployable sensors to detect these dangerous microbes.

628278 Synthetic Biology Mediated Electrochemical Sensing Strategy

Abstract:

Synthetic biology enables integration of biomolecules with different functions in a modular manner toward desired applications. However, the output signal of current biochemical circuits primarily remains optical. We herein present strategies to apply electrochemistry to evaluate the performance of multi-function biochemical circuits. A CRISPR mediated primer-exchange-reaction based genetic circuit is designed toward probing specific regions of the genome of SARS-CoV-2. The output of the gene circuit is connected with a single-use, microfabricated gold sensing surface, enabling the direct translation of biomolecular information into electrical signal. The integration of electrochemistry into synthetic biology can lead to a new paradigm for sensor development and can also be a high-throughput interface for the investigation of large-scale gene circuits.

631801 Hydrogel-Encapsulated Gold Nanoshells Prepared By Inverse Emulsion Polymerization As a Biosensor for Sjögren's Syndrome Protein Markers

Abstract:

Proteomics research has led to identification of increasing numbers of differences in the protein makeup of biological fluids, such as saliva, tears, urine, or serum, due to the onset of disease; however, more work is needed to improve the molecular recognition capabilities of existing biosensors before these markers can see more widespread adoption in clinical practice. One approach for more robust, low-cost recognition elements is synthetic receptors based on polymeric hydrogels, but this approach is challenged by the lack of specificity toward the target analytes, which reduces their viability for recognition in biological fluids due to a myriad of competing molecules. One solution is differential sensor arrays that exploit semi-selective recognitive elements in order to screen for differences in multiple protein biomarkers in a single assay. In this work, we explore the capabilities of using inverse emulsion polymerization to form libraries of multifunctional hydrogel networks on the surface of gold nanoshells (AuNS) as semi-selective recognition elements for use in a localized surface plasmon resonance-based biosensor for the detection of biomarkers associated with Sjögren's syndrome.

Hydrogel nanoparticles were prepared by an inverse emulsion copolymerization of methacrylic acid with acrylamide or *N*-isopropylacrylamide, crosslinked with *N,N'*-methylene bisacrylamide. The particles were characterized by dynamic light scattering, potentiometric titration, and Fourier-transform infrared spectroscopy. AuNS were prepared using the method of seeded growth of gold colloids on aminated silica nanoparticles followed by encapsulation with poly(maleic anhydride-*alt*-1-octadecene)-g-poly(ethylene glycol) methacrylate (PMAO-g-PEGMA) graft copolymer to improve particle stability. Hydrogel-coated AuNS were prepared with the same procedure as the nanoparticle synthesis with the addition of PMAO-g-PEGMA AuNS to the aqueous phase of the inverse emulsion. UV-Vis spectroscopy was used to measure the shifts in the wavelength of the localized surface plasmon resonance for the hydrogel-coated AuNS. Inverse emulsion synthesis was found to lead to thin surface coatings for both formulations and shows immense promise for the creation of a library of hydrogels with various functional groups that could be used to alter affinity for different protein biomarkers from patient biological fluids.

Work supported by National Institutes of Health Grant R01-EB022025

628700 Silk-Based Microneedle Biosensor for Sustainable Food Supply Chain

Abstract:

Manipulation of hierarchical assembly of biopolymers enables utilization of their functionalities for engineering purposes. Silk fibroin extracted *Bombyx mori* cocoons is an emerging technical material for food contact applications due to its non-toxic and edible nature. In this study, we present a silk-based microneedle injection system for food bacteria and spoilage detection. We developed internal pore structures of microneedles through the water-annealing process leading to increased beta-sheet formation. The microneedles delivered internal food fluid via capillary action in pores to a colorimetric sensor on the needle array's backside. Within a day of injection, we identified *E. coli* contamination and common food spoilage in fish fillets by color changes of the polydiacetylene-based sensor. We also showed that the microneedle sensor could pierce commercial food packaging, implying a successful adaptation of the technology downstream in conventional food supply chains. Therefore, this study provides insights into global food safety and food waste reduction.

Engineered CRISPR-ENHANCE system for clinical detection of SARS-CoV-2 RNA

Long T. Nguyen, Santosh R. Rananaware, Brianna L.M. Pizzano, Brandon T. Stone, Piyush K. Jain*

*Corresponding and presenting author

The Severe Acute Respiratory Syndrome-Coronavirus-2 (SARS-CoV-2) has resulted into 2019 novel coronavirus disease (COVID-19) pandemic with confirmed almost 3 million deaths and over 136 million confirmed cases worldwide as of April 12th 2021. There is an urgent need for the development of a rapid, sensitive, specific, inexpensive, easy-to-use, and accessible testing kit for SARS-CoV-2 that can be implemented as a point-of-care diagnostic and a home-based testing kit. Most Class 2 CRISPR/Cas systems, including CRISPR/Cas12 (Type V), CRISPR/Cas14 (Type V), and CRISPR/Cas 13 (Type VI) mediate a nonspecific collateral trans-cleavage of random DNA and RNA after binding or cis-cleavage of their target DNA or RNA. For CRISPR/Cas12a, this multiple-turnover trans-cleavage activity is only initiated once the crRNA/Cas12a complex is bound to its target ssDNA or dsDNA that acts as an activator. This trans-cleavage activity has been widely exploited for nucleic acid detection and has been combined with fluorescence-based, paper-based, and electrochemical-based sensing technologies to develop rapid and sensitive diagnostics. Therefore, Type V and VI CRISPR/Cas systems have recently emerged as diagnostics for detecting nucleic acids including SARS-CoV-2 RNA.[1-11] However, current CRISPR/Cas12a systems are limited to a picomolar detection limit unless the target is either pre-amplified in some manner. The Cas12a-based DETECTR technology from Mammoth Biosciences and Cas13a-based SHERLOCK technology from Sherlock Biosciences have been approved by the FDA under EUA as lab-based diagnostics for detecting SARS-CoV-2 genomic RNA in a fluorescence-based assay.

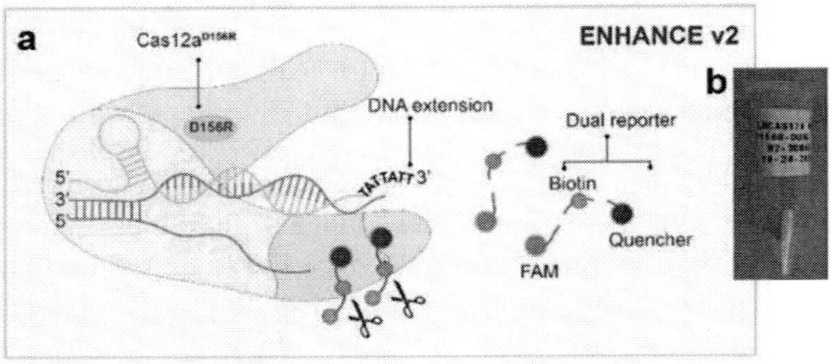

Figure 1: *Development and validation of a lyophilized and lateral flow version ENHANCEv2. (a) ENHANCEv2 is comprised of an engineered LbCas12a with D156R mutation, crRNAs containing 7-nt DNA extensions on 3' end, and a dual reporter. Once a dsDNA target is recognized by ENHANCEv2, it cleaves dual reporters and allows for enhanced multimodal detection using a lateral flow assay and a fluorescence-based assay in a single assay. (b) A lyophilized version of ENHANCEv2 demonstrated enhanced stability at room temperature for at least 30 days.*

We recently discovered that crRNAs can tolerate various length extensions with DNA or RNA on both 3'- and 5'-ends, while the phosphorothioate extensions reduced or eliminated the trans-cleavage activity (Nguyen et al., *Nature Communications*, 2020)[9]. In particular, an engineered crRNA containing a 7-nt DNA extensions on its 3'-end, drastically increased the trans-cleavage activity of CRISPR/Cas12a and increased specificity of detection. By extending the 3'- or 5'-ends of the crRNA with different lengths of DNA, RNA, and phosphorothioate DNA, we discovered a new self-catalytic behavior and an augmented rate of Cas12a-mediated trans-cleavage activity as high as 3.5-fold compared to the wild type crRNA, making it the fastest reported CRISPR/Cas12a in terms of trans-cleavage activity, and termed it as CRISPR-ENHANCE. This in turn reflected an unprecedented improvement in sensitivity and limit of detection of nucleic acid targets down to the femtomolar level for a variety of clinically-relevant nucleic acid targets including prostate cancer antigen 3 (PCA3), HIV, HCV, and SARS-CoV-2 RNA without target pre-amplification within 30 minutes. By combining with a reverse transcription-loop-mediated isothermal amplification (RT-LAMP) step, 100 aM of SARS-CoV-2 RNA could be detected using a lateral flow assay within 60 minutes (Nguyen et al., *Methods*, 2021)[10].

By further optimization of crRNAs and combining with a target pre-amplification strategy using a reverse transcription-loop-mediated isothermal amplification (RT-LAMP) step, a single molecule detection of SARS-CoV-2 RNA in nasopharyngeal swabs was achieved within 50 minutes. Compared to the wild-type crRNAs, our CRISPR-ENHANCE system demonstrated up to 5-fold higher sensitivity in detecting SARS-CoV-2 RNA. By combining engineered crRNAs with a Cas12a mutant (D156R), a multimodal reporter, and lyophilizing the components, the test could be conducted within 33 minutes using both a fluorescence-based assay and a lateral flow assay. The CRISPR-ENHANCE demonstrated high inclusivity and exclusivity and achieved 96.7% sensitivity and 96.7% specificity in clinical samples and was stable for at least 30 days at room temperature (Figure 1, Nguyen et al., *MedRxiv*, 2021- *in revision at Communications Medicine-Nature*)[11].

References

1. Broughton JP, Deng X, Yu G, et al. CRISPR-Cas12-based detection of SARS-CoV-2. *Nat Biotechnol.* Apr 2020;doi:10.1038/s41587-020-0513-4
2. Abudayyeh OO, Gootenberg JS, Kellner MJ, Zhang F. Nucleic Acid Detection of Plant Genes Using CRISPR-Cas13. *CRISPR J.* Jun 2019;2:165-171. doi:10.1089/crispr.2019.0011
3. Kellner MJ, Koob JG, Gootenberg JS, Abudayyeh OO, Zhang F. SHERLOCK: nucleic acid detection with CRISPR nucleases. *Nat Protoc.* Oct 2019;14(10):2986-3012. doi:10.1038/s41596-019-0210-2
4. Gootenberg JS, Abudayyeh OO, Kellner MJ, Joung J, Collins JJ, Zhang F. Multiplexed and portable nucleic acid detection platform with Cas13, Cas12a, and Csm6. *Science.* 04 2018;360(6387):439-444. doi:10.1126/science.aaq0179
5. Myhrvold C, Freije CA, Gootenberg JS, et al. Field-deployable viral diagnostics using CRISPR-Cas13. *Science.* 04 2018;360(6387):444-448. doi:10.1126/science.aas8836
6. Gootenberg JS, Abudayyeh OO, Lee JW, et al. Nucleic acid detection with CRISPR-Cas13a/C2c2. *Science.* 04 2017;356(6336):438-442. doi:10.1126/science.aam9321
7. Aquino-Jarquin G. CRISPR-Cas14 is now part of the artillery for gene editing and molecular diagnostic. *Nanomedicine.* 06 2019;18:428-431. doi:10.1016/j.nano.2019.03.006
8. Chen JS, Ma E, Harrington LB, et al. CRISPR-Cas12a target binding unleashes indiscriminate single-stranded DNase activity. *Science.* 04 2018;360(6387):436-439. doi:10.1126/science.aar6245
9. Nguyen LT, Smith BM, Jain PK. Enhancement of trans-cleavage activity of Cas12a with engineered crRNA enables amplified nucleic acid detection. *bioRxiv.* 2020:2020.04.13.036079. doi:10.1101/2020.04.13.036079
10. Nguyen LT, Gurijala J, Rananaware SR, Pizzano BLM, Stone BT, Jain PK. CRISPR-ENHANCE: An enhanced nucleic acid detection platform using Cas12a. *Methods.* Feb 2021;doi:10.1016/j.ymeth.2021.02.001
11. Nguyen LT, Rananaware SR, Pizzano BLM, Stone BT, Jain PK. Engineered CRISPR/Cas12a Enables Rapid SARS-CoV-2 Detection. *MedRxiv.* 2020;doi:10.1101/2020.12.23.20248725

633118 Computational Design of MOF-Based Electronic Noses for Disease Detection By Breath

Abstract:

Although it well-known that our breath contains a multitude of biomarkers for various diseases, breath-based screenings and diagnostics are almost non-existent in a clinical setting. This is largely due to the fact the current state-of the-art gas sensing technologies, namely gas chromatography-mass spectrometry (GCMS), are too bulky and too expensive for screening a large number of samples. Furthermore, due to the sheer number of compounds in breath, simple low-cost sensors such as mass-based and chemiresistive devices struggle to differentiate between the compound(s) of interest and other unknown gases. Thus, conventional diagnostics such as blood tests are still preferred. However, researchers have envisioned using gas sensing arrays, often called electronic noses, to achieve a level of sensitivity and selectivity comparable to GCMS, while maintaining the cost, speed, and size advantages of simple sensors, ultimately creating a device which would make breath-based screening and diagnostics competitive with conventional techniques.

Herein, we discuss the use of metal-organic frameworks (MOFs) in electronic noses of surface acoustic wave (SAW) devices for breath-based disease detection. Owing to the number of MOFs (>130,000) one could choose from, as well as the number of gases found in our breath (700+), the task of building an electronic nose is as much of a big data problem as it is an engineering challenge. Thus, to properly address this big data aspect of the problem, our efforts to design an electronic nose have been computational in nature. Using Grand Canonical Monte Carlo (GCMC) simulations, we can generate massive libraries of adsorption data which can be used to match against real sensor outputs, enabling quantification of a gas mixture. However, as we desire to include more gases and detect them in smaller concentrations, the number of gas mixtures which would need to be simulated grows exponentially. Thus, in order to maintain a reasonable computational demand, we pivoted to a new model in which we use GCMC simulations to generate a library of adsorption coefficients which enable us to determine the adsorbed mass of trace gas species without running a distinct simulation for each new set of concentrations. In turn, we created a flexible algorithm for the quantification of trace gas species. Since most of the gases of interest in breath are present in trace quantities, the above method works very well for simultaneously quantifying numerous compounds in breath, and using that information for disease screening and diagnostics.

629784 Interaction of Acetylcholinesterase with Phorate

Abstract:

Detection of residual organophosphates in food is of paramount importance in current scenario. Biosensors for pesticide detection may be developed on solid surfaces by monitoring the interaction of acetylcholinesterase with pesticides. Acetylcholinesterase is a hydrolytic enzyme which hydrolyzes acetylcholine, a neurotransmitter, into choline and acetate. This enzyme gets inhibited in the presence of pesticides. Utilizing this property, protein based sensor on solid surfaces may be developed. The interaction studies were carried out on silica surface using acetylcholinesterase and phorate, an organophosphate. Self-assembled monolayer was created by immobilization of enzyme on silica surface and further pesticide is allowed to interact with adsorbed protein. Binding of immobilized protein and phorate on surface was evaluated in terms of the mass and thickness change of surface using quartz crystal microbalance which is calculated from frequency and dissipation factor change using Sauerbrey equation. Low value of dissipation factor (ΔD) corresponds to stronger binding of pesticide with protein. The nature of chemical reaction between acetylcholinesterase and phorate was analyzed by the stoichiometry and energy released. The interaction study was carried out at low phorate concentration of 5 ppm of phortae. These findings helps in the development of protein based pesticide detection sensor so as to provide user friendly technique for detection in raw food samples.

Keywords: Acetylcholinesterase, ITC, pesticide, QCM.

References:

1. Singh, A. K. *et al.*, 1999 Development of sensors for direct detection of organophosphates. Part I: immobilization, characterization and stabilization of acetylcholinesterase and organophosphate hydrolase on silica supports. Biosensors & Bioelectronics. 14:703–713.

627267 Hyperspectral Imaging Sensors Enhance on-Line Food Quality and Authenticity Inspection.

Abstract:

Used extensively as a Remote Sensing technique in Agriculture and Environmental monitoring, Hyperspectral Imaging (HSI) makes rapid advances in industrial process monitoring and shows great potential to become a valuable technique for rapid qualitative and quantitative assessment of food quality, safety, and authenticity.

Hyperspectral Imaging is a technique that collects spectral information, resolved over 100+ bands, for every pixel in an image. This wealth of spectral and spatial information acquired during a scan can be used to solve challenges like detection of foreign objects or discrimination between types of objects for various applications.

Modern HSI cameras and integrated systems, coupled together with image processing and statistical tools, can be used for product or process characterization and for delivering quantified results.

Significant value can be realized by using HSI as a tool for early inspection at receiving stations in food processing facilities. Here the ability to quickly scan large quantities of product and get results without the need to prepare samples for laboratory analysis can help the user reduce the cost of quality. Better informed decisions can be made early in the process, helping the user assign value to a specific batch or decide that a certain batch does not meet quality criteria. The information from an early inspection can also be used to help improve quality upstream – in the fields, groves and farms.

An example of this type of HSI application is prediction of sweetness or Brix content in oranges. Brix value is one of the most important quality parameters for growers and processors and traditionally requires a sample to be sent to a lab, oranges to be juiced and soluble solids content measured using a refractometer. A case study described in this presentation shows a prediction of Brix based on HSI scan of a few oranges. Refractometer data was used as a reference. Results of the prediction suggest that the technique could be used for quality assessment of samples under controlled conditions. Methodology used to develop a spectral classification model also helped in creating an optimized model for run-time spectral classification. This enabled the system to produce real time Brix results for batches of oranges passing under the scanner.

Another area where HSI delivers significant value is food fraud detection. The technique is well established in forensic applications and art authentication but now, with reduced costs, simplified user interfaces and AI powered algorithms, HSI becomes a usable tool for industrial QC. Food fraud is a major global problem. In particular, products that move through global supply chains and are sold in powder or small flake form are susceptible to foul play. Oregano, for example can be adulterated with cheaper ingredients like myrtle or olive leaf. Detecting this visually is quite difficult and processing facilities often resort to tedious, manual inspection under a microscope or sending samples for a chemical analysis. With HSI the user can quickly scan a sample of the product and develop a model that would differentiate a pure sample from an adulterated one. Further image processing can help in producing quantified results as in % purity, but even a simple color classification can help make a quick decision if the sample is good or if it requires a more detailed analysis.

An important enabler for Hyperspectral technology in industrial applications is the software for real time spectral classification. Machine Learning algorithms allow even a non-specialist user to create practical, deployable models in a matter of hours. Important progress is also being made in sensor design, taking advantage of increasing power of embedded image processing computers complete hyperspectral systems can now be packaged into compact, rugged housings making the technology useful as a process analytical tool. Embedded computing and on-board processing or raw hyperspectral data overcomes one if the major historical hurdles for adoption of HSI – enormous quantities of raw data generated by the cameras. Modern integrated HSI systems with embedded processing can perform run-time spectral classification and produce an output that consists of a waterfall image stream, object metadata or predicted quality attributes like moisture content, Brix and other results of regression modeling.

processing technology in the Hyperspec MV.X is removing the complexity associated with handling vast amounts of raw data. By performing spectral classification and post processing images on board the device we can output actionable results – object locations, size or predicted values directly to the end user.

This solution produces a complete, fully integrated system in a single ingress protected housing. The design lends itself well to being integrated with other automation or operated as a standalone quality monitoring system. Networking capabilities of embedded computing turn it into a smart device that can be deployed as part of a digital solutions strategy. The scalability of having the same system suitable for both small scale bench testing and for full on-line production environment is a big value add for the users.

621789 Collagen Type II Quantification with Pd@Pt Nanoparticle-Linked Immunosorbent Assay

Abstract:

In this presentation, we show a method for quantitatively detecting a larger-sized analyte, Collagen type II, using a Pd@Pt nanoparticle-linked immunosorbent assay.

Extracellular matrix (ECM), including cartilage tissue, is made up of proteoglycan and collagens, which act as the main structural proteins. Collagen types I, II, and III are essential fibril-forming collagens that comprise 80 – 90% of all collagens found in the human body. Collagen type II makes up approximately 50% of the protein and 85 – 90% of collagen in articular cartilage. Articular cartilage is the tissue that connects diarthrodial joints, providing a cushion between the bones.

The Centers for Disease Control and Prevention (CDC) estimates that over 32.5 million US adults suffer from arthritis. The most common type of arthritis, osteoarthritis, is a degenerative joint disease frequently found in the hands, hips, and knees. Osteoarthritis causes symptoms such as pain, swelling, and less flexibility because of the erosion and damage of articular cartilage. When osteoarthritis is present in a joint, components of the ECM, Collage type II and aggrecan, are broken down by the enzyme collagenase. Therefore, Collage type II quantification is a potential method for observing the degree of cartilage damage in osteoarthritis patients.

There presently is no treatment or cure for osteoarthritis, which is why we are studying articular cartilage regeneration through the use of scaffold-based techniques. Collagen type II is one component of the regenerative medical scaffolding used for cartilage and tissue engineering applications. Therefore, the ability to produce Collage type II and quantify the production is needed.

Nanozymes have enzyme-like catalytic activity like that of peroxidase and catalase and finds usage in various field because nanoparticles have the capacity to be modified, functionalized, and mass produced at low cost, and are inherently stable. In this study we used mesoporous palladium@platinum (Pd@Pt) nanoparticles a nanozyme with peroxidase-like catalytic activity that consists of a palladium core with a platinum shell. In addition, these nanoparticles are highly stable with respect to pH and temperature, allow simple conjugation with biological components such as antibodies, and allow simple separation methods through centrifugation. These characteristics allow us to use Pd@Pt nanoparticles as detection agents instead of the enzymes in immunosorbent assays.

Since Collage type II is large enough to have two epitopes, a sandwich immunoassay was used to capture and label this larger analyte with two different antibodies. Assay conditions were optimized for Collage type II quantification using a Pd@Pt nanoparticle-linked immunosorbent assay. An optimal concentration of coating Abs was found to be 2.5 μg/mL at a pH of 8.4 – 8.6, Ab conjugated to Pd@Pt NPs at a level of 7.5 μg Ab to 1 mg/mL Pd@Pt, and a concentration of Ab – nanoparticle conjugates of 2 mg/mL. We plan to apply our sandwich detection method to engineered tissue samples. Our use of this method broadens the scope of Pd@Pt nanoparticle-linked immunosorbent assays beyond our previous study for small analytes such as pesticides. We expect this assay to be adopted for use in the fields of bioengineering, tissue engineering, and bioanalysis.

629941 Cellular Hitchhiking for Targeting Nanoparticles

Abstract:

Nanoparticle-based drug delivery systems are widely explored to improve the biological outcome of chemo and immunotherapy. However, poor vascular circulation, limited targeting, and the inability to negotiate many biological barriers are key hurdles in their clinical translation. Biology has provided many examples of successful "carriers" in the form of circulatory cells, which routinely overcome the hurdles faced by synthetic nanoparticle systems. Our laboratory has explored blood-cell inspired drug delivery systems that take advantage of the abilities of red blood cells and macrophages. We have explored "cellular hitchhiking" which involves combining synthetic particles with circulatory cells to drastically alter the in vivo fate of the synthetic particles. I will provide an overview of the principles and two examples of hitchhiking-based nanoparticle delivery.

623632 Nanomedicine for Improved Ocular Drug Delivery

Abstract:

Eye drops are the dominant dosage form for the ocular route, and are used for a wide range of indications, including glaucoma, dry eye, inflammation, infection and allergy. However, as with all exposed epithelial surfaces, there are innate protective mechanisms that hinder efficient drug delivery to the eye. Thus, only a small fraction of drugs administered by eye drops is typically absorbed into the eye, often necessitating multiple doses per day. As the number of eye drop doses per day increases, so does the potential for ocular surface irritation and systemic side effects, while patient compliance to the treatment regimen decreases. To address these issues, we have developed hypotonic gelling formulations that provide enhanced topical ocular drug delivery for a range of drugs for different ocular indications affecting both the anterior and posterior segment. We demonstrate that our hypotonic, lower concentration thermogelling formulations outperform both conventional isotonic thermogelling formulations and commercial eye drops for drugs that lower intraocular pressure, both hydrophilic (brimonidine tartrate) and hydrophobic (brinzolamide), as well as a hydrophobic peptide drug (cyclosporine) used for treating dry eye. Importantly, our hypotonic gelling formulation caused no signs of ocular irritation or toxicity with twice-daily administration for up to five weeks in rabbits, as measured by corneal staining, blink-rate test and corneal histology. Further, characterization of the uniformity, refractive index and absorbance suggest that visual clarity is maintained and the shear-thinning properties are similar to those of commercial lubricating eye drops. We also provide data to support the hypothesis that by holding the drug at high concentration at the ocular surface, drug delivery to the posterior retina and choroid can be increased. We demonstrate proof-of-principle with topically delivering drugs that inhibit hypoxia inducible factor 1 (acriflavine) and vascular endothelial growth factor receptor (sunitinib) to prevent laser-induced choroidal neovascularization in mice. Importantly, therapeutic drug delivery to the posterior segment was confirmed in large animals with eyes similar in size to the human eye, also providing protection against prevention of laser-induced choroidal neovascularization. Our findings highlight the importance of formulations that conform to the ocular surface before viscosity enhancement for increased and prolonged ocular surface contact and drug absorption. The therapeutic implications span a variety of ocular indications with the potential for enhanced disease management and patient quality of life.

626528 Nanoparticle Physiochemical Design Features to Modulate Pulmonary Innate Immune Cell Response

Abstract:

While respiratory diseases globally number among the top causes of mortality, the field of pulmonary drug delivery has lagged behind other routes of administration in the application of novel therapeutic approaches. Most current inhalers are employed in the treatment of asthma and have yet to effectively address many significant respiratory conditions. A significant need remains to expand inhaled drug delivery beyond the limited number of current therapeutics. Thus, the overall goal of the Fromen lab is to develop personalized aerosol immunomodulatory therapeutics for patients with a wide range of airway conditions. To overcome current challenges in the field, our group designs new approaches to therapeutic pulmonary aerosols, leveraging chemical engineering principles, immune engineering approaches, and additive manufacturing.

In this talk, I will discuss recent work from our lab to create and characterize nanoparticle therapeutics for efficient pulmonary vaccines and therapeutics. We have explored a range of new particle chemistries, including polymeric and metal organic framework (MOF) platforms, and have evaluated how various physiochemical properties, including particle degradation rate, influence innate immune responses. Interestingly, we find that the lifespan of macrophages, innate immune cells that protect the deepest regions of the lung, are highly regulated by phagocytosis of particles through anti-apoptotic and non-inflammatory mechanisms. We find that nanoparticle (NP) dosing and cellular internalization via a *single* phagocytosis event significantly enhances survival of *ex vivo* cultures of primary bone marrow-derived, alveolar, and peritoneal macrophages over particle-free controls. The enhanced survival is attributed to suppression of caspase-dependent apoptosis and is linked to phagocytosis and lysosomal signaling, which was also found to occur *in vivo* following two separate routes of administration. We find this phenomenon is applicable to NPs of alternative chemistries and varies as a function of dosing frequency and particle degradation rate, indicating the potential universality of this phenomenon with relevant drug delivery particles and the significance of using NP therapeutics to regulate cell lifespan. Overall, our efforts will continue to advance aerosol design rules to inform personalized inhaled particulate formulations, with potential future applications for novel treatments for cancer, inflammation, vaccination, and allergy.

622270 Thermomechanical Properties of Nanodiamond Superstructures in Interlayer-Bonded Twisted Bilayer Graphene

Abstract:

Graphene derivatives and graphene-based metamaterials and nanocomposites have been studied extensively due to their exceptional properties. These properties can be tuned precisely by tailoring structural features introduced in these 2D materials by means of chemical functionalization and defect engineering. Toward this end, here, we report results of a systematic computational study of a new class of graphene-based nanomaterials, namely, nanodiamond superstructures embedded in interlayer-bonded twisted bilayer graphene (IB-TBG). As a result of interlayer bonding within twisted bilayer graphene, induced by chemical functionalization due to patterned hydrogenation, superstructures of nanodiamond domains are formed between the two graphene layers. We have found that such IB-TBG nanodiamond superstructures with an interlayer bond density higher than a critical value undergo ductile failure upon uniaxial tensile straining, which makes these structures very appealing toward developing mechanically superior graphene-based metamaterials.

Based on molecular-dynamics (MD) simulations according to a reliable interatomic bond-order potential, we determine a broad range of thermomechanical properties of IB-TBG nanodiamond superstructures. Specifically, we study the mechanical response of these superstructures to indentation loading and shear straining based on MD simulations of nanoindentation tests and dynamic shear straining tests. We establish the dependences of the elastic modulus, maximum stress sustained upon indentation, and shear strength of the superstructures on the full range of their structural parameters, especially on the diamond fraction in the nanocomposite superstructure as measured by the concentration of sp^3-hybridized interlayer-bonded C atoms, f_{sp3}. The resulting structural responses and fracture mechanisms of the superstructures under nanoindentation testing and shear straining also are characterized in detail. We have found that superstructures characterized by a less-than-critical f_{sp3} exhibit a strongly nonlinear inelastic response to indentation up to the onset of fracture. A non-dissipative and non-recoverable stiffening effect caused by the relative twisting between the top and the bottom layer of the nanodiamond clusters in the central region of the indented superstructure is found to induce a kink-like discontinuity and oscillations in the corresponding force-displacement indentation loading curve. In addition, we have found that the 2D elastic modulus and the deformability of this type of superstructure, as measured by the vertical deflection at the breaking point, are relatively insensitive to the interlayer bond density, while the maximum sustained stress decreases monotonically with increasing f_{sp3} but remains very high even at high f_{sp3}.

Finally, we report results for the lattice thermal conductivity of these superstructures based on non-equilibrium MD (NEMD) simulations of thermal transport. We find that the lattice thermal conductivity is reduced significantly with increasing diamond fraction in these nanocomposite superstructures, which also makes these 2D carbon nanomaterials very promising for thermal management applications.

629751 Interactions of Graphene Oxide Nanosheets with Blood-Related Entities and Their Implications for Hematological Disorders

Abstract:

Blood-related disorders are some of the most fundamental and complex abnormalities as they can cause critical health conditions, including heart attack, stroke, and cancer [1, 2]. Although nanomaterials have been actively explored for various biomedical applications [3-5], they have not been really investigated for hematological applications. Among all classes of nanomaterials explored for biomedicine, the two-dimensional (2D) graphene oxide (GO) nanosheets stand out because of their unique and versatile physicochemical properties [6], which may be beneficial for hematological applications [7-10]. Herein, we examined the interactions of GO nanosheets with different blood-related entities, particularly plasma proteins, blood cells, and blood-borne parasites, to unravel the biophysicochemical activities of GO nanosheets for hematological applications. First, we evaluated the GO-protein associations, specifically the adsorption, equilibrium binding, and conformational stability of plasma proteins upon interacting with GO nanosheets [7, 8]. The interactions of GO nanosheets with whole blood under flow condition, and with *Plasmodium falciparum* malaria parasites, were then probed to assess the antithrombotic and antimalarial properties of GO nanosheets [9, 10]. GO nanosheets were noted to possess high loading capacity for major blood plasma proteins and the nanosheets did not induce protein denaturation. These interactions were protein-specific and might be considerably influenced by the lateral size distribution and concentration of GO nanosheets. We also observed that functionalized GO nanosheets displayed antithrombotic characteristic under flow condition. Moreover, pristine GO nanosheets, irrespective of their size distribution, exhibited antimalarial activity by mitigating the malaria parasite invasion on red blood cells and delaying the parasite maturation. Overall, with their high biomolecule loading capacity and unique antithrombotic and antimalarial characteristics, GO nanosheets are highly promising for hematological applications. This study will provide a deeper insight into nanomaterial-blood interactions and potentially facilitate further explorations into the formulation of nanotechnology-based strategies for biomedical applications.

References

[1] Furie et al., *The New England Journal of Medicine* 359 (2008), 938-949.

[2] Burns et al., *The Lancet Haematology* 3 (2016), E362-E370.

[3] Doane et al., *Chemical Society Reviews* 41 (2012), 2885-2911.

[4] Chen et al., *Chemical Reviews* 116 (2016), 2826-2885.

[5] Shi et al., *Nature Reviews Cancer* 17 (2017), 20-37.

[6] Kenry et al., *Biomaterials* 155 (2018), 236-250.

[7] Kenry et al., *NPG Asia Materials* 9 (2017), e422.

[8] Kenry et al., *Nanoscale* 8 (2016), 9425-944[illegible].

[9] Kenry et al., *Small* 11 (2015), 5105-5117.

[10] Kenry et al., *Nanoscale* 9 (2017), 14065-14073.

622804 Sustainable Production of Graphene from Petroleum Coke Using Electrochemical Exfoliation

Abstract:

Petroleum coke is a solid, carbonaceous by-product of oil refining and is normally used for heating or as an anode in aluminum and steel production. These applications contribute to carbon emissions, but here we show that petroleum coke has another potential avenue: as a precursor for graphene production. This path presents an environmentally and economically sustainable use for a low-value industrial stream. Electrochemical exfoliation is used to produce graphene nanosheets from petroleum coke, rather than graphite. The final product is separated from the unreacted material by a two-step centrifuging process. SEM and TEM images confirm that the final product contains few-layered nanosheets, and the Raman spectra confirm that the exfoliated coke product is indeed graphene. Post-annealing of this product substantially increases the electrical conductivity. This new finding holds potential for the petroleum industry to produce a value-added nanomaterial and enhance the economic impact of slurry oil and slurry oil-derived coke streams by orders of magnitude; this route also allows these streams to be directed away from high-emissions uses.

632769 Interleukin 6 (IL-6) Targeted Corona Phase Molecular Recognition Using Fluorescent Nanosensors

Abstract:

Molecular recognition sites that specifically binds with a target molecule are essential for clinical research, disease diagnosis, and therapeutic development. Corona Phase Molecular Recognition (CoPhMoRe) is an emerging technique that uses amphiphilic polymer adsorbed onto a nanoparticle surface to generate a synthetic recognition site. To date, CoPhMoRe has demonstrated its versatility by successfully detecting molecules including nitric oxide, hydrogen peroxide, vitamins, neurotransmitters, carbohydrates, proteins, steroid, and small molecule pharmaceutical drugs using polymer wrapped single walled carbon nanotubes (SWNTs). However, this method has not yet been widely extended to macromolecular analytes. Herein, we interface near-infrared fluorescent SWNTs with a panel of amphiphilic polymers varying in monomer composition and length, and use them against a library of inflammatory biomolecules, revealing a specific corona phase that recognizes interleukin 6 (IL-6) with high selectivity. The IL-6 recognition shows dose-dependent responses, and exhibited excellent stability with response from 10 ng/mL to 100 μg/mL. Two-stage binding of IL-6 with the SWNT corona is suggested by sensor dynamic studies. The IL-6 sensor is also deposited onto intravascular catheter and displays functionality in biological environment. These results open new pathways for synthetic molecular recognition that detects biological macromolecules, and hold great promise for medical and clinical applications.

627960 Understanding the Impact of Pore-Polymer Interactions on the Mobility of Poly(ethyleneimine) Confined in Mesoporous SBA-15: Quasi-Elastic Neutron Scattering Studies

Abstract:

Supported amines are considered a promising class of CO_2 sorbents based on their capability of capturing CO_2 from various sources such as flue gas stream from power plants (5-15% CO_2) as well as ambient air (~400 ppm). Among supported amines, poly(amines) physically impregnated within the mesopores of solid oxide supports have been extensively used for achieving high amine contents (i.e., high capacity) and physical/chemical stability (i.e., limited loss via evaporation/oxidative degradation of amine compounds).[1] Within supported amine systems, the physical properties of the poly(amines) play key roles in determining CO_2 uptake performance. In particular, the chain dynamics of poly(amines) govern diffusion rates of CO_2, which ultimately determine CO_2 sorption/desorption dynamics.[2] Therefore, understanding the dynamics of poly(amine) species confined in solid supports will promote designing optimal supported amines for CO_2 capture.

Herein, we study the impact of the interaction modes between the pore surface and poly(amines) with a model system comprised of branched-poly(ethyleneimine) (BPEI) supported in mesoporous SBA-15 silica. Chain mobilities of BPEI are quantitatively analyzed using quasi-elastic neutron scattering (QENS) in tandem with coarse-grained molecular dynamics (MD) simulations. The pore-polymer interactions are tuned by grafting organosilanes having different end groups to the silica surface, leading to different levels of attraction between BPEI and the pore walls of silica. Chain mobilities of BPEI undergoing four different interaction modes with pore walls are investigated. The first mode brings covalent tethering of amine sites of BPEI to pore surface via halide-amine reaction, where alkyl halides are grafted onto pore surfaces. The second interaction is acid-base and hydrogen bonding interactions based on silanols decorating the native SBA-15 silica, which strongly interact with amine groups of BPEI. The third interaction considers relatively weak interaction with pore surface and BPEI by substituting acidic silanols into basic aminoalkyl groups on the pore walls, so that pore walls interact with BPEI mainly by inter-amino hydrogen bonding. Lastly, repulsive, van der Waals interaction (via grafting aliphatic chains on pore walls) yields the weakest interaction between pore walls and BPEI. QENS (High-Flux Backscattering Spectrometer; HFBS at NIST) and MD simulation revealed that the chain mobilities of BPEI are significantly influenced by different surface groups appearing on the pore walls. The strong/transient interaction (acid-base, hydrogen bonding) due to silanol-BPEI interaction yielded the slowest motions, covalent bonding and repulsive interactions resulted in slightly increased mobility, and hydrogen bonding (aminoalkyl-BPEI; without acid-base interaction) brought significantly recovered mobility as compared to bulk PEI. We hypothesize that PEI's mobility depends not only on the attraction level with the pore walls but also on its chain conformation in the presence of surface capping groups on the pore walls. To understand further, these results are currently being coupled with additional QENS experiments, alongside insights developed via other techniques.

References:

(1) Choi, S.; Drese, J. H.; Jones, C. W. Adsorbent Materials for Carbon Dioxide Capture from Large Anthropogenic Point Sources. *ChemSusChem* **2009**, *2* (9), 796–854.

(2) Holewinski, A.; Sakwa-Novak, M. A.; Carrillo, J.-M. Y.; Potter, M. E.; Ellebracht, N.; Rother, G.; Sumpter, B. G.; Jones, C. W. Aminopolymer Mobility and Support Interactions in Silica-PEI Composites for CO2 Capture Applications: A Quasielastic Neutron Scattering Study. *J. Phys. Chem. B* **2017**, *121* (27), 6721–6731.

622996 Overcoming the Permeability-Selectivity Tradeoff Via the Design of Heterogeneous Pore Wall Chemistries

Abstract:

Developing next-generation membranes for water treatment demands system-specific materials with precisely tuned functionality. In particular, the purification and reuse of highly contaminated water calls for solute-specific designs where a key challenge is the separation of small neutral solutes. One such solute is boric acid, which is strictly regulated in water reuse for agricultural irrigation. Precise tuning of membrane pore chemistry requires a fundamental understanding of the mechanism by which functional moieties at the pore wall modulate the transport of water and solutes. Simulations enable the nanoscopic investigation of the mechanism by which a chemically heterogeneous pore modulates hydration water and thus its water-mediated interactions with solute molecules. Here, we use molecular dynamics simulations to investigate water and boric acid structure, dynamics, and transport in a cylindrical nanopore patterned with nonpolar methyl and polar hydroxyl groups. We find that changes in the arrangement of the methyl and hydroxyl groups, in addition to the density of each group, affect the transport of both water and boric acid in the pore. These changes not only affect the selectivity of water over boric acid, but also the permeability of water through the pore, leading to improvements in the permeability-selectivity behavior, a pernicious trade-off in membrane technology. We thus apply an optimization procedure based on a genetic algorithm to design the chemical patterning of the pore wall to simultaneously optimize water selectivity and permeability. The optimization identifies non-intuitive patterns that significantly hinder the transport of boric acid through the pore, simply by altering the functional group patterning. This inverse design procedure and the accompanying molecular-level understanding of the effect of chemical heterogeneity on solute selectivity in a pore will inform the design of more efficient water purification membranes, as well as the modulation of water-mediated surface-solute interactions in other systems.

624568 Hydrogen Bond Patterns of Deep Eutectic Solvents and Non-Deep Eutectic Solvents

Abstract:

Deep eutectic solvents (DESs) have shown promise as sustainable solvents to extract substrates from aqueous solutions as they are composed of natural compounds. Currently, a trial-and-error method is used to find new DESs experimentally. This process is inefficient and time consuming as all permutations of binary systems must be tested and then ruled out if no DES is formed. Hydrogen bond interactions between the binary compounds are considered the prevailing reason for the eutectic point. We investigated the relationship between hydrogen bond features and the potential of forming DESs for a binary system using a computational approach. We conducted molecular dynamics simulations to analyze the structural and dynamic properties of hydrogen bonds in 50 DESs and 100 non-DES systems and investigated their pattern using a statistical approach. Our results could be used to develop molecular principles for designing new DES based on natural hydrogen bond donors and acceptors.

620632 Silica-Kaolinite Interface for Shale Gas Application: A Molecular Modeling Study

Abstract:

Methane mostly makes up the constituent of shale gas and it is currently being exploited in fulfilling the world's energy demands. Molecular modeling techniques including Density Functional Theory (DFT) and Molecular Dynamics (MD) techniques are employed to understand methane transport in the pores at typical reservoir conditions. Herein, we study the interfacial phenomenon of methane adsorption on Silica-Kaolinite interface. This interface is created as it simulates a shale model which in reality is made up of clay (kaolinite) and quartz-like (silica) materials. The simulations revealed that the interface is formed by a chemical bond between a silicon atom from the silica surface and two oxygen atoms from the kaolinite surface. The adsorption of methane was studied at three different positions on the Silica-Kaolinite interface namely, silica-dominated region, interface region, and kaolinite-dominated region. The mode of adsorption of methane irrespective of its position on the interface was found to be physisorption with the kaolinite region having a greater adsorption. This is in agreement to previous studies which used experimental techniques for study methane adsorption. The results found herein which will be useful in shale estimating the original gas in place for shale reservoir application.

This work is funded by NPRP grant 12S-0130-190023 from from the Qatar National Research Fund (a member of Qatar Foundation). The statements made herein are solely the responsibility of the authors.

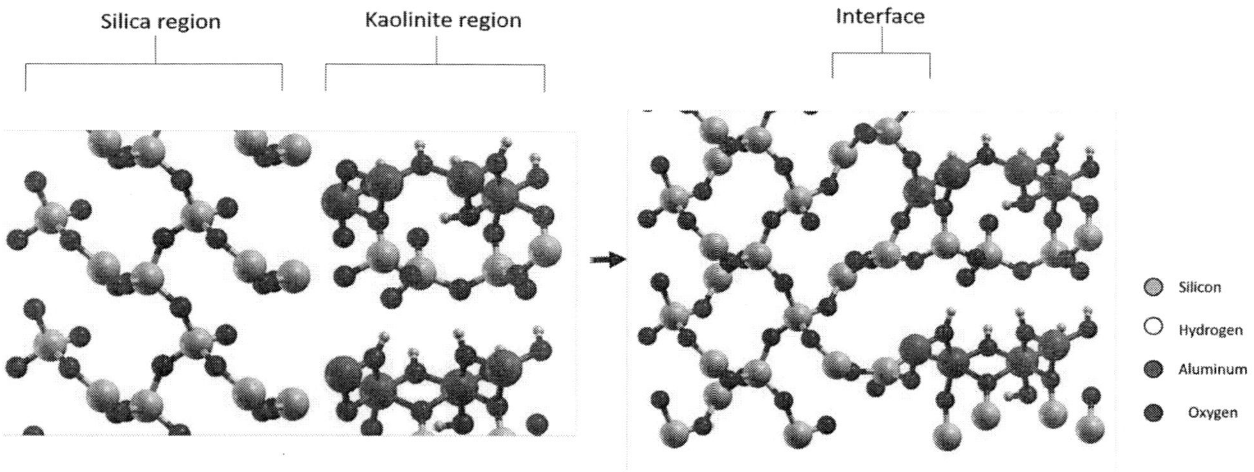

629486 Investigation on the Sodiation of Selenium-Graphene for Na-Ion Batteries: A First-Principles Study

Abstract:

Sodium–selenium (Na–Se) batteries have recently attracted attention as an emerging alternative for the traditional Li-ion batteries owing to their high theoretical volumetric capacity (~ 3250 mAh/cm^3) and electronic conductivity of Se (~ 10^{-3} S/m). Carbon composites are proposed as promising host materials to buffer the volume fluctuation of Se during charge/discharge, provide conductive channels, and physically trap the high-order polyselenides. In this talk, we evaluate the role of carbon host in establishing the reaction kinetics as the fundamental understanding of the sodiation of the Se-C composite is still limited. Based on first-principles simulations, we first present interfacial structure, charge transfer, and diffusion characteristics of Na atoms at the interface between Se and graphene (C). Our calculations demonstrate that there is a substantial enrichment of Na atoms at the interface and corresponding electron transfer to both Se and C nearby. The interactive strength of Se with graphene is found to increase with sodiation, which could lead to mechanical failures as compressive stress build-up at the interface. The presence of C host results in a more populated distribution of longer Se chains, as less reduced Se (due to electron transfer toward graphene) tends to be clustered. We observed a faster mobility of interfacial Na atoms along the graphene surface when compared to the diffusion of Na atoms in Se. Our calculations also show that the Na–Se/C system exhibits n-type metallic characteristic, which could provide fast electron flux for the sodiation reaction.

630087 Molecular Insights into the Electrowetting Behavior of Aqueous Ionic Liquid

Abstract:

Electro-wetting is the modulation in the wettability of a surface prompted by an external voltage [1, 2] across the solid/liquid interface. Ionic liquids (ILs) which have unique physicochemical properties,[3] gain industrial significance, due to its development as a promising alternative medium for electrowetting. This phenomenon is exploited in various applications such as variable fluid lens systems, electrowetting displays, programmable optical filters, paint drying, micro motors, electronic microreactors, liquid optics, or lab-on-a-chip devices.[4-7]

Molecular dynamics simulations were applied to study the wetting behavior of nanoscale aqueous hydrophilic and hydrophobic Imidazolium-based ionic liquid (IL) droplets on a solid graphite substrate subjected to the perpendicular electric field. Imminent transformation in the droplet configuration was observed at $E = 0.08$ V/Å both for hydrophobic ILs [EMIM][NTF_2] and SPC/E water droplets. However, for hydrophilic, [EMIM][BF_4], the droplet was completely stretched to be column-shaped under the electric field $E = 0.09$ V/Å for lower weight percentages of ILs and $E > 0.1$ V/Å at a higher weight percentage of ILs (> 30 wt%), because of the hydrophilic nature of [EMIM][BF_4]. We explore the impact of electric-field through various parameters such as mass density distribution, contact angle, orientation, and hydrogen bonds.

[1] A. Quinn, R. Sedev, J. Ralston, Contact Angle Saturation in Electrowetting, The Journal of Physical Chemistry B, 109 (2005) 6268-6275.

[2] J. Restolho, J.L. Mata, B. Saramago, Electrowetting of ionic liquids: Contact angle saturation and irreversibility, The Journal of Physical Chemistry C, 113 (2009) 9321-9327.

[3] J.T. Hirvi, T.A. Pakkanen, Wetting of Nanogrooved Polymer Surfaces, Langmuir, 23 (2007) 7724-7729.

[4] D. Bratko, C.D. Daub, K. Leung, A. Luzar, Effect of field direction on electrowetting in a nanopore, Journal of the American Chemical Society, 129 (2007) 2504-2510.

[5] Y.S. Nanayakkara, H. Moon, D.W. Armstrong, A tunable ionic liquid based rc filter using electrowetting: A new concept, ACS Applied Materials & Interfaces, 2 (2010) 1785-1787.

[6] Y.S. Nanayakkara, H. Moon, T. Payagala, A.B. Wijeratne, J.A. Crank, P.S. Sharma, D.W. Armstrong, A fundamental study on electrowetting by traditional and multifunctional ionic liquids: possible use in electrowetting on dielectric-based microfluidic applications, Analytical chemistry, 80 (2008) 7690-7698.

[7] Y. Wang, G.A. Voth, Unique spatial heterogeneity in ionic liquids, Journal of the American Chemical Society, 127 (2005) 12192-12193.

626423 Biotemplating of Barley Stripe Mosaic Virus Virus-like Particles for Directed Synthesis of Metal Nanomaterials

Abstract:

The synthesis of nanoscale materials of uniform morphology and well-defined architecture is of a substantial interest recently. Biotemplating has become an emerging field in which natural biomolecular objects are utilized for creating functional, hierarchical, controlled patterned structures with nanometric precision. Especially, viral biotemplating has shown great potential in applications because of the features of precise dimension, diversity of architecture and the amenability to genetic/chemical engineering. Recently, in order to overcome the restrictions associated with viral genome and plant-based production, a novel virus biotemplate, Barley stripe mosaic virus virus-like particle (BSMV-VLP) is designed and engineered from an alternative genetically tractable bacterial-based expression platform to produce BSMV capsid proteins which expedites the processing duration for large-scale mass production.

Our most recent focus is on the directed synthesis of metal nanomaterials with the developed BSMV-VLP biotemplates. The hydrothermal synthesis yielded continuous and uniform palladium coatings on the BSMV-VLPs without an exogenous reducing agent. The VLP-mediated nanorods are with a wide range of length scales, of greater uniformity and higher coverage than the ones synthesized with in planta-produced BSMV virion. In addition, it further expands the biotemplating capability of BSMV-VLP to create bimetallic complex nanomaterials. In the absence of an external agent, the synthesis presented the exceptional capability to mineralize different metal nanoparticles onto the virion and form binary metallic mixtures of palladium-platinum, palladium-gold and palladium-copper. The synthesized nanorods of uniform and dense coatings were imaged with transmission electron microscopy (TEM). Scanning transmission electron microscopy (STEM), energy dispersive X-ray spectroscopy (EDS) and X-ray photoelectron spectroscopy (XPS) were used for elemental composition characterization. We've demonstrated a controllable and effective approach for metallic nanorod synthesis on the engineered BSMV-VLP biotemplate using the hydrothermal solution processing. Taken together, this synthetic approach has explored the synthetic palette and opened up enormous possibilities in the bottom-up nanofabrication of versatile and tunable organic-inorganic nanoscaled complex and would facilitate future engineering industrial applications.

632164 Metal-Organic Frameworks for Vaccine Stabilization: A Translational and Mechanistic Study

Abstract:

The global share of vaccinated infants is nearly 90%, yet in the poorest 70 countries, less than 10% of 1-year-olds are fully vaccinated. This discrepancy is largely due to failures in the "cold chain," which necessitates the uninterrupted refrigeration of vaccines from the moment of manufacture until administration. Cold chain failures result in significant vaccine losses globally, posing a humanitarian problem as well as an economic one: cold-chain logistics costs are estimated to reach $16.6B this year.

One interdisciplinary solution to the cold chain problem relies on Metal-organic frameworks (MOFs)--a class of highly tunable porous materials. Recently, some groups have shown that encapsulating biological macromolecules (e.g., proteins and viruses) within a MOF scaffold can impart heightened stability to guest molecules far beyond ambient temperatures, perhaps enabling vaccine storage and transport without refrigeration.

In this work, we report the synthesis and thermostability of Tetanus Toxoid encapsulated in a Zeolitic Imidazolate Framework (ZIF-8). The toxoid protein retains exceptional immunological activity upon release from the framework after aging at 60°C for 1 month (shown by ELISA). Because the current literature focuses primarily on model guests of little clinical interest, our work represents an important contribution as it provides evidence of improved stability by MOF encapsulation on a clinically relevant (and never-before reported) guest protein.

While this research is predominantly translational, we also utilize crystallographic, spectroscopic, and scattering techniques to unveil the mechanism of guest stabilization. Through a collaboration with the Advanced Photon Source at Argonne National Labs, we have analyzed these biocomposites with Small-Angle X-ray Scattering (SAXS) generated by a synchrotron radiation source. With our approach of scaled spectra subtraction, we are able to deduce structural properties (i.e., size and shape) of encapsulated guests while they are being heated *in situ*. These experiments have provided novel insights into the heightened stability afforded by MOF biocomposites, which appears to rely on the tight confinement of guest molecules within MOF cavities (thus preventing deformation and aggregation of proteins).

622748 DNA-Caged Polymer Micelles for Cell and Tissue Labeling

Abstract:

Purpose: This research leverages DNA nanotechnology (i.e., DNA tiles) to create 'erasable' labeling technologies for pathology imaging. Multiplexed images allow for the labeling of multiple targets simultaneously with high image resolution, which means that complicated pathways in samples such as tumors can be studied. However, the gold standard for pathology is single biomarker analysis in sequential slides, which precludes simultaneous evaluation of multiple signaling markers in the same cell. In conventional immunocyto- and immunohisto- chemistry processes, cells or tissue are stained with antibodies conjugated to fluorophores or colored dye reporter molecules designed for single use. However, one path forward is the use of erasable labels that would enable the same sample to be interrogated multiple times. The methods used in erasing fluorescent labels often involve harsh chemicals or photobleaching with UV light. Both of these processes have the potential to damage delicate structures found in thin tissue slices. Another type of erasable labeling uses single-stranded (ss)DNA. The ssDNA, usually labeled with a fluorophore, is attached to an antibody. Depending upon the attachment scheme, the fluorophore can then be photobleached or the entire strand can be removed by a strand displacement. This method is problematic in that the dyes employed are not protected from the environment and if only a single fluorophore or dye molecule is used, the signal may be weak. Here, we show the use of DNA-caged micelles for erasable pathology labeling. This system is based on gentle DNA dehybridization interactions that minimize tissue damage. Additionally, the dyes used in cell and tissue labeling are protected by the micelle coating.

Methods: DNA tiles are used as a surface coating for polymer nanocomposites based on electrostatic interactions, as proposed by Kurokawa et al.[1]. The DNA tiles were modified to include fewer interlocking strands in favor of unique targeting strands that bind the labeling antibody. The targeting strands allow for reversible attachment of DNA caged particles to free ssDNA or to ssDNA-modified antibodies. Solution erase was demonstrated by the addition ssDNA tagged with fluorophores that were removed via strand displacement reactions.

Results and Implications: The first step is proving existence and formation of DNA cages. TEM images of micelles, DNA tiles, and DNA caged micelles were taken. Each image shows a different structure and particle size, established the formation of DNA cages. Next, to ensure that optimal structure formation and surface binding, FRET and saturation studies were conducted. FRET between dye on the micelle surface and quencher on the DNA shows that DNA does bind to the surface of the micelles rather than forming separate populations in solution. Additionally, DNA tile structures were designed to show fluorescence quenching when the repeating units interlock, showing cage formation rather than random surface absorption. Saturation studies show that there is a maximum amount of DNA that can bind to the micelle surface. We have also observed that DNA cages are stable in common buffers used such as 5% Triton X100 or Tween-20. When the DNA cages are used in solution erasing as described above, up to 85% of the signal can be repeatability erased in 8 to 15 minutes. Since the action of erasing uses a solution of ssDNA in PBS it is designed to be gentle on tissue. We then proved this hypothesis correct with the use of DNA cages and erasing solution in a variety of tissue samples with no evidence of damage to the samples. Finally, when DNA cages are used as a fluorescent reporter, up to 80% of the signal is erased after 20 minutes of exposure to ssDNA.

Our results indicated DNA cage formation on the nanocomposite surface and proof of concept labeling in cells and tissues. Proper diagnosis and treatment, especially of cancer, is highly dependent on cell and tissue labeling. DNA caged particles allow erasable labeling without harsh treatments. These approaches could be applied to label the same target in multiple cycles (e.g., super-resolution imaging) or enable different targets to be labeled sequentially. When coupled with advanced image analysis, this approach could be used to create 3D images. Therefore, DNA cages have the potential to significantly change cell and tissue labeling.

Reference

1 Kurokawa, C. *et al.* DNA cytoskeleton for stabilizing artificial cells. *Proc Natl Acad Sci U S A* **114**, 7228-7233, doi:10.1073/pnas.1702208114 (2017).

629651 DNA Origami Tubes with Reconfigurable Cross-Sections

Abstract:

DNA nanotechnology has emerged over the last decades as a promising approach for various biological, mechanical, and environmental applications. The programmable nature of DNA bases provides a framework to construct nanoscale devices with complex geometry, precise motion, and controlled dynamic and mechanical properties. Dynamic devices in particular have emerged as promising tools for various biological, mechanical, and environmental applications such as biophysical measurement devices, nanorobotic components, triggered drug delivery vehicles, and nanobiosensors. The unique properties of DNA nanostructures also make them highly attractive for bottom-up assembly of materials with dynamically programmable properties. Hence, coupling across length scales, i.e. using individual devices on the molecular scale to make larger assemblies including microscale robotic systems and materials, is an exciting avenue for DNA nanotechnology. There have been major advances over the last decade in higher order assembly of DNA structures, including recent efforts to make reconfigurable arrays and lattices; however, the dynamic behavior of these assemblies remains limited compared to the major advances in complexity of individual devices. Here our goal is to provide an approach using dynamic DNA devices capable of reconfigurable higher order assembly to advance the complexity of reconfigurable arrays and lattices on a larger scale.

We leverage scaffolded DNA origami (DO), which is a molecular self-assembly approach where many short single-stranded DNA (ssDNA) oligos, referred to as "staples," bind with a long ssDNA strand of a known sequence called a "scaffold." The staples bind to the scaffold using Watson-Crick complementary base-pairing in precise locations to create double-stranded DNA (dsDNA) regions that fold into precisely designed shapes. Early work on DO structures focused on static structures and shapes, whereas more recent work on dynamic DO structures include motion or actuation to increase complexity. Advanced 3D dynamic devices include objects with 1D motion such as hinges or pistons, and mechanisms with more complex motion such as crank-sliders or Bennet linkages with programmed kinematic motion or actuation capabilities. A number of recent efforts have expanded DO to larger scales through hierarchical assemblies that integrate many individual structures in a controlled way, and prior efforts have demonstrated integration of reconfigurable constructs into arrays. However, the reconfiguration of higher order assemblies remains quite simple to 1D rotation and only open-to-close reconfiguration. Therefore, we developed a reconfigurable 6-component DNA origami mechanism, the 6-bar, that can be reconfigured into multiple shapes (Figure 1A) and assembled into linear arrays (Figure 1C) with different cross-sections. Our goal is to demonstrate the reconfiguration of DO shapes into a variety of stiff cross-sections and demonstrating the reconfiguration of DO tubes between several distinct cross-section shapes.

The 6-bar consists of 6 arms or links, consisting of 12 double-stranded DNA (dsDNA) helices arranged in a 3x4 square lattice cross section. These 6 bundles are connected via ssDNA hinges to form a closed-loop mechanism. The "free" state is flexible and contains 172 nucleotide (nt) scaffold loops on each bundle. These scaffold loops are present to enable the formation of struts that connect neighboring components at a specific angle to determine the shape of the mechanism. The struts can create bundle angles of 60°, 90°, 120°, and 180° (Figure 1A). Each folded configuration contains three full struts, and all the staples comprising the struts contain overhanging ssDNA extensions to serve as toe-holds for toe-hold mediated strand displacement (TMSD). Each configuration can thus reconfigure back to the free (i.e. no strut) state with the use of TMSD, and then to a different geometry through the addition of new strut strands. Figure 1B illustrates a transition between an open rectangle to a flat closed geometry with high efficiency of reconfiguration. Reconfiguring the individual 6-bar devices provides a basis for reconfiguring higher-order assemblies. We expand the reconfiguration to polymer "tubes" with the same reconfigurable cross-sections. We can assemble these DO polymers into several different cross-section shapes (Figure 1C). These polymers allow us to bridge the gap between individual device reconfiguration to multi-device reconfiguration.

In addition to demonstrating reconfiguration of individual devices and tubes, the 6-bar provides an interesting platform to study free energy differences of the various conformations that only vary by the composition of the struts. Introducing different sets of strut staples to the free (i.e no strut) state allows us to analyze the preferred, or stable, configurations. Exhibiting reconfiguration between multi-states also allows us to study the energetics between the open to closed loop structures. We expanded this idea to structural tuning of the interactions as an approach to control formation of higher order assemblies such as tubes with varying cross-sections across the length. Studying the energetics of the higher order assemblies may provide new ways to control assembly to create more complex reconfigurable DO structures.

This work expands upon molecular motors and machines to include nanoscale and micron-scale mechanisms. The 6-bar mechanism and 6-bar polymers could be used to organize components over a micron length-scale or create assemblies with inhomogeneous properties. DNA origami includes many applications such as drug delivery, signal transmission, and biological mechanisms, however has expanded broadly to many different fields outside of biology. Using DNA as a material to construct mechanisms studied in fields such as chemical, mechanical, and bio-engineering has been an amazing advancement provided the biological nature and application of DNA.

Figure 1: **A.** (left) Course-grain molecular dynamic simulations of 6-bar mechanism. (right) Transmission electron microscope (TEM) image averages of each 6-bar configuration (scale bar = 50nm). **B.** (left) Reconfiguration schematic. (right) TEM images at each step of reconfiguration with image average insets (scale bar = 50nm). **C** (top) 6-bar polymer schematics and (bottom) TEM images of the rectangular and compact configurations (scale bar = 100nm).

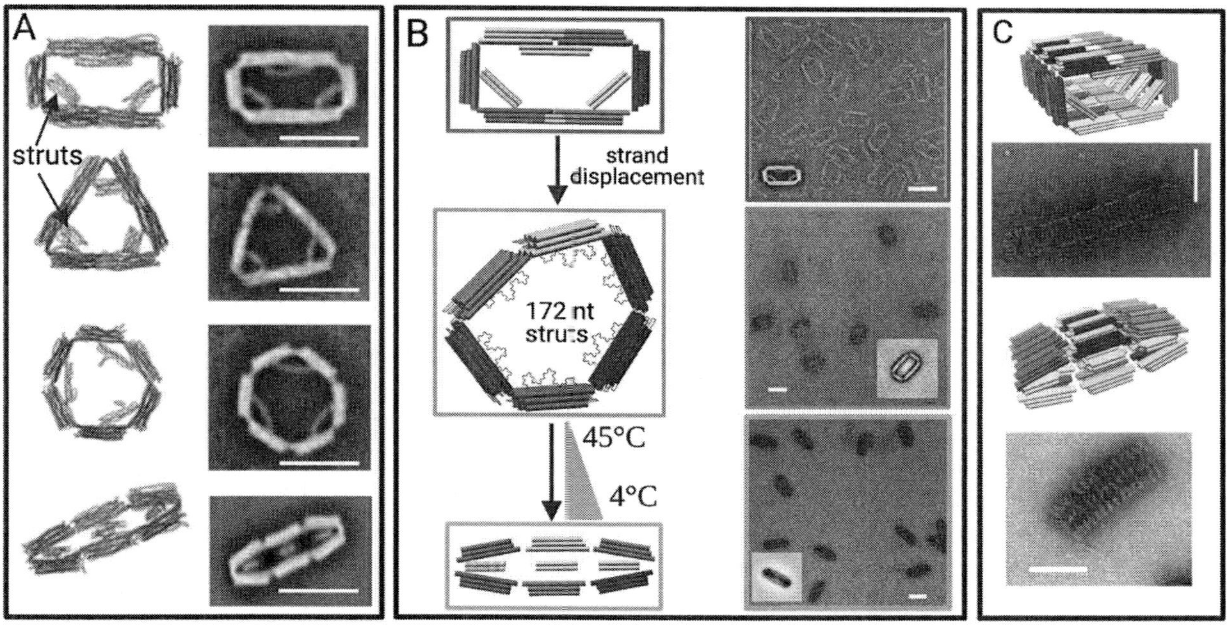

623869 CRISPR/Cas-Directed Hierarchical Assembly of Protein-DNA Hybrid Nanostructures

Abstract:

Nanomaterials that incorporate both DNA and proteins are of particular interest in the field of bionanotechnology since they combine the highly precise, programmable assembly characteristics of Watson-Crick pairing with the chemical heterogeneity and bioactivity of proteins.[1] These hybrid structures are integral to the development of new methods of drug delivery, biosensing, and the study of fundamental characteristics and behaviors of biomolecules. Specifically, the 3D spatial control of DNA origami scaffolds enables the construction of complex shapes such as nanocages or nanotweezers. Current methods of interfacing proteins and DNA have focused on direct covalent modification of the protein of interest or fusion to independent DNA-binding protein motifs. However, the former is limited by the need for multiple reaction steps which may damage the protein, and the latter is limited by the need to change the primary amino acid sequence to recognize each distinct, corresponding DNA binding site.[2] A new alternative interfacing technique utilizes CRISPR-associated (Cas) nucleases, a novel class of DNA-binding proteins that possess high binding affinity (K_D ~0.05-0.5nM) and versatile binding specificity. Recent work by Lim et. al. exploits these characteristics for the spatial organization of multi-enzyme complexes on a linear double stranded DNA scaffold.[3]

Here, we employ a strategy that builds upon prior CRISPR/Cas-mediated assembly techniques to connect multiple distinct protein nanofibers to 3D DNA origami templates in a modular and programmable way. The protein nanofibers are comprised of -prefoldin (PFD) oligomers derived from the hyperthermophile *Methanocaldococcus jannaschii,* exhibit remarkable thermostability (T_M of 93 ^{o}C), and have been shown to be excellent scaffolds for the immobilization of enzymes and metalloproteins to enhance catalytic activity and generate conductive nanowires.[3,4] We utilize the SpyCatcher-SpyTag conjugation system to connect these fibers to a catalytically inactive variant of Cas9 nuclease derived from *Streptococcus pyogenes* (dCas9) that can recognize and bind to a target DNA sequence without generating a double stranded break. To accomplish this, the engineered capping protein TERM (a mutant of PFD) is genetically fused to the SpyCatcher domain (TERM-SpyCatcher) while the 13-amino acid SpyTag peptide is expressed at the C-terminus of dCas9 (dCas9-SpyTag). When combined, TERM-SpyCatcher associates selectively with the terminus of PFD filaments. Subsequent co-incubation of this product with dCas9-SpyTag generates dCas9-capped nanofibers (dCas9-fibers). For the 3D DNA origami structures (cuboids, triangles, etc.), we incorporate one to four modified staple strands which have an additional sequence of 53 base pairs extending perpendicular to each face that function as attachment points. Through the addition of dCas9-fibers to one of four distinct, corresponding sequences of single-stranded guide RNA (sgRNA) we are able to control where on the structure the fibers attach. We propose this hierarchical assembly approach can be used to generate hybrid protein-DNA nanostructures that can be further functionalized with distinct recognition moieties (*e.g.* enzymes, peptides, nanoparticles), and examples of such systems will be discussed.

632998 In-Situ Monitoring with a Surface Plasmonic Enhanced Native Fluorescence in the Ultraviolet Spectral Region

Abstract:

Fluorescence imaging has been widespread use in clinical diagnosis and monitoring in biological systems. Almost all biological molecules present some level of intrinsic fluorescence in the ultraviolet (UV) region of the spectrum. UV spectroscopy of biomolecules performed in the 250–400 nm range has fluorescence emission peaks at UV wavelengths depending on their molecular structure. Enhancement of intrinsic emission of proteins and other biomolecules can be achieved through metal surface plasmon resonance (SPR) in the UV region, contributing to label-free detection. Despite its ability to achieve label-free molecularly recognition and quantification, the applications of UV plasmonics for biosensing have been limited due to challenges in engineering (nanostructure design, optimization, and fabrication) and materials science (detailed composition analysis). The signal level of intrinsic fluorescence is still orders of magnitude less than that of organic dye molecules in a fluorescence-labeled assay.

We report the development of novel surface plasmon resonance-based aluminum nanostructure with a microfluidic platform for rapid characterization of enhanced fluorescence from tryptophan using UV spectrometer. The detection mechanism is label-free, enhances fluorescence intensity by plasmonic structure, and able to detect molecules simultaneously. A microfluidic flow cell has been fabricated on an aluminum nano-hole arrays substrate to enable UV measurement of target molecules. UV spectroscopy performed from tryptophan shows an increasing fluorescence emission trend with concentration. The limit of detection is calculated to be ~20 fmol which is about 20 times more sensitive than aluminum thin film substrate. Our results have demonstrated label-free, sensitive, and in-situ detection of target molecules by combining microfluidics and ultraviolet plasmonic nanostructures.

622351 Data Management Schema Design for Effective Nanoparticle Formulation for Probing and Treating Neurological Disease

Abstract:

Background: Nanomedicine has limited translation from pre-clinical research to clinical application for non-cancer neurological disease [1]. To improve translation, it is important to understand the dynamic interactions between nanomedicine platforms and the complex brain environment. To investigate dynamic nanoparticle-brain interactions with a range of nano-based delivery systems, it is useful to develop a database that can organize and integrate key experimental variables and factors that might influence the effectiveness of nanotherapeutics in the brain.

Database management systems already commonly exist in clinical medical applications. For example, The Brain Database is a clinical neuroanatomy database developed over three decades ago [2]. However, pre-clinical data management schemas have different complexities than clinical applications. Pre-clinical databases need management schemas that logically connect experimental methodologies, which can be highly repetitive, vary in quality, are prone to rapid iteration and evolution, and often research lab or facility-specific. There is limited or non-existent literature on effective database management for nanoformulations used preclinically, but similar approaches for utilizing databases exist in computational cell biology for managing data-rich biological cell process data [3]. For example, BioNumbers is a computational cell biology database that stores and organizes key quantitative features from cell biology [4]. The BioNumbers database is composed mainly via natural language processing of the literature, containing one specific number per property. In contrast, nanoformulation-brain databases are focused on organizing wet lab experimental research prior to the publication phase and often have repetitive experiments with varying values for each property. A successful standardized data management schema for pre-clinical nanotherapeutic experimental pipelines has three main effects: (1) increased sustainability for data organization and storage, (2) increased insight into related variables and their interactions, and (3) increased searchability of specific results or methods.

Methods:

To show the efficacy of a well-designed pre-clinical database management schema for nanoformulation data, we create a database using PostgreSQL that performs three functions: (1) Obtains the nanoparticle formulation variables that produce specific nanoparticle sizes and charges; (2) Connects nanoformulations to all related experimental elements, including nanoformulation methodology, biological specimen prep, and biological specimen characterization; and (3) Defines a standard lab protocol for regularly updating the database as new nanoformulation experiments are carried out.

Our first step was to design the entity-relationship diagram to relate three main aspects: experiment, nanoformulation, and biological specimen (e.g. application). We then created a process flow diagram as a visual representation of data management from the chemical engineering, experimentalist perspective (Figure 1). This process flow diagram was converted into an entity-relationship diagram using LucidChart. With LucidChart, we were able to visualize our relationship details within the database, including all variables, keys, and variable types. From LucidChart, we directly exported the database schema in PostgreSQL and subsequently imported the data onto a local server. We hosted the database locally due to the overall data size (51.5 kilobytes). Before we uploaded all of our data to the schema, we cleaned and standardized the data sets, which were collected from nine independent researchers.

After the database was created, we developed standard queries for accessing nanoformulation and biological data based on commonly used variables. For the nanoformulation data, we use three key physical characteristics of nanoformulations to inform our queries: size, zeta potential, and polydispersity index (PDI). Nanoparticle size and surface charge can influence nanoparticle passage across the blood-brain barrier and penetrate within the brain parenchyma [5]. Zeta potential is an indirect measure of nanoparticle surface charge and PDI is an indicator of particle size uniformity. We developed standardized queries which allow us to search biological characterization information including catalase activity, cytotoxicity over time, and glutathione levels for experiments that use nanoformulations in biological specimen. Finally, we developed a standardized method of inputting data into a cloud-based database by creating group spreadsheets with commonly used variables for each nanoformulation technique and characterization method. We additionally standardized a method for assigning unique keys for each individual nanoformulation, researchers, characterization experiment, and biological specimen based on input order and time information.

Results:

Once the database was set up, data input, and queries written, we evaluated our data management schema's effectiveness via two methods. The first approach evaluated the times it took for each query to run and return results. The first set of our queries selects all nanoparticle formulations with specific characterization features in either size, zeta potential, polydispersity, or a mix of these three variables. The second set of our queries obtains all nanoparticle formulations and their formulation variables for specific nanoparticle features for either single emulsion, double emulsion, or nanoprecipitation methodologies. The slowest query ran with an average time of 72.5 milliseconds. Before creating the database, to obtain a list of nanoformulation methodologies and their variables for a specific nanoformulation characterization result, an interested researcher would need to make personal asks to other researchers or perform additional nanoformulation experiments. Both options are time-consuming and ineffective. With a standard database, it is possible for researchers to easily access previous experiments from other researchers in an efficient way by searching directly for a variable of interest.

The second evaluation method visualizes our nanoformulations' physical characteristics for each query. The visualization of our nanoformulation data is important because it allows us to check that the queries are doing what we intended and provides valuable scientific insight into our formulation data. For example, if we query the size, zeta potential, and PDI, it is clear that the queries successfully minimize our data set to those within user-defined parameters. Our first query restricted the size of formulated nanoparticles to within 50 to 100 nanometers. This query restricted the total results from 538 formulations to 237. Our second query added an additional restriction on the zeta potential values to reside between -10 and 10 millivolts. Query 2 restricted our results from 237 formulations to 160. Finally, we restricted our overall formulations with an additional parameter on PDI values between 0.0 and 0.2 which brought our viable nanoformulation with these parameters to 109. These evaluation plots also provide unique insight into how our different nanoformulation methods compare to one another. For example, we can now query our data for specific nanoparticle feature ranges and then compare our single emulsion methodologies to our double emulsion methodologies. We are also able to gain insight from our zeta potential chart as is. From queries 2 and 3, we learned that although our nanoformulations are within the neutral range (-10 to 10 mV), they still tend towards negative values. The database works to successfully relate formulations and methodologies with specific formulation outcomes.

Conclusions:

Developing and utilizing a database management schema on already established experimental pipelines provided new insights, effective data storage, and data querying options. We were able to query our nanoformulations to obtain approximately 100 nanoformulation batches that fit the requirements imposed in this project. We also have shown a proof-of-concept that our experimental data can benefit from a well-structured database management schema. Additionally, we are currently expanding the database to match nanoformulation experiments to biological experiments. We are doing this by creating additional tables for biological specimen variables and related biological characterization methods. For example, an *in vivo* rat experiment can be connected to behavioral testing results, litter information, and any tissue processing. Once biological information is input into the database, we use an experiment tag to connect nanoformulations and their characterizations to the biological specimen they were tested in. The database organizes these connections for efficient querying, and then we develop queries for nanoformulations based on biological features such as species, age, or behavioral results.

Finally, we plan to develop standard practices for discussing, planning, and participating in better data management. These practices will lead to better data utilization and data longevity, and greater reproducibility and interpretability of datasets generated from independent researchers. Insights into relationships and trends in nanomedicine formulations have benefitted from developing this database management schema. A standard data management schema for experimental nanoformulation experiments decreases the need for repetitive experimentation, connects nanoformulation variables with biological outcome in an interpretable and efficient way, and enables science-informed querying of any standardized variables. The overall benefits of a standardized database contribute to efficient experimentation and cross-experimental insight, which will ultimately improve pre-clinical to clinical translation of nanoformulations.

References:

[1] Anselmo, A.C. and S. Mitragotri, *Nanoparticles in the clinic: An update.* Bioengineering & Translational Medicine, 2019. **4**(3).

[2] Wetheim, S., *The Brain Database: A Multimedia Neuroscience Database for Research and Teaching.* 1989.

[3] Loew, L.M. and J.C. Schaff, *The Virtual Cell: a software environment for computational cell biology.* Trends in Biotechnology, 2001. **19**(10): p. 401-406.

[4] Milo, R., et al., *BioNumbers—the database of key numbers in molecular and cell biology.* Nucleic Acids Research, 2010. **38**(suppl_1): p. D750-D753.

[5] Helmbrecht, H., et al., *Governing transport principles for nanotherapeutic application in the brain.* Current Opinion in Chemical Engineering, 2020. **30**: p. 112-119.

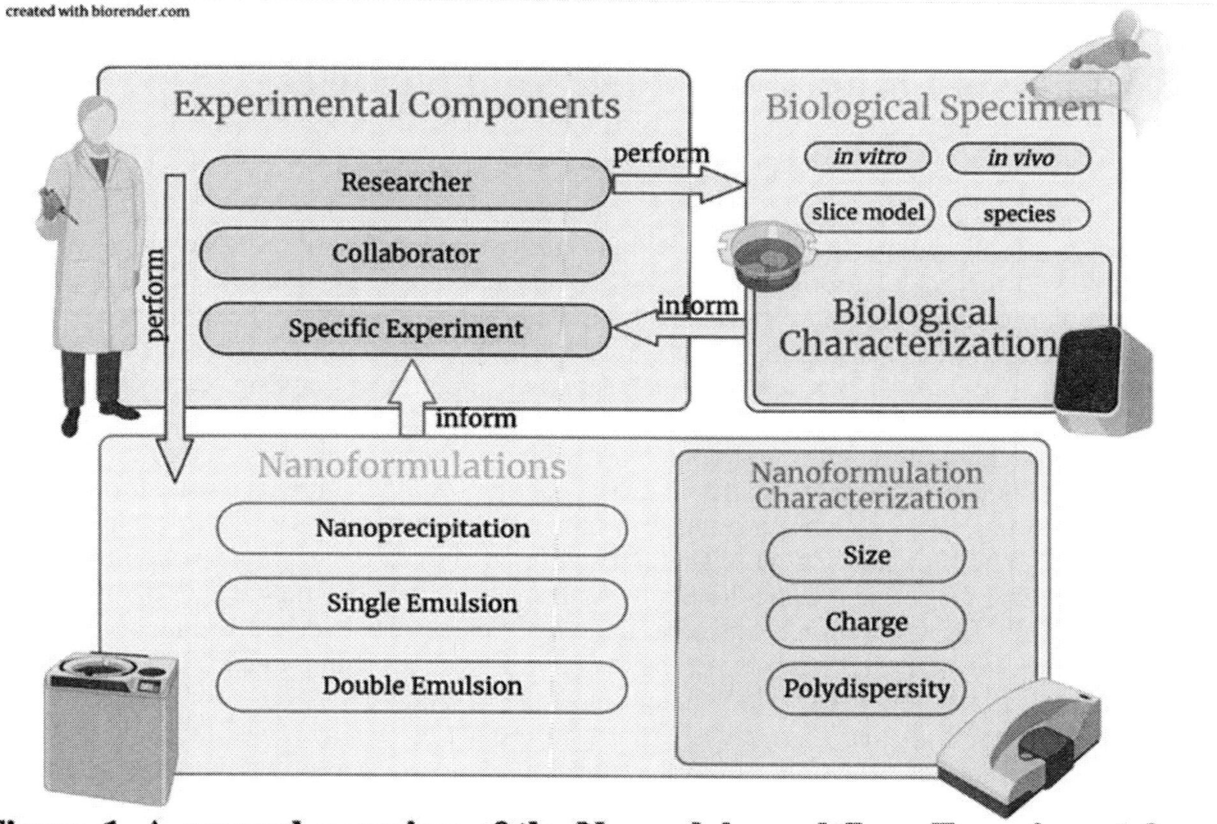

Figure 1. A general overview of the Nance lab workflow. Experimental components include specific experiment details, researcher, and collaborator details. Biological details include the specimen: *in vitro, in vivo,* slice model, and species alongside the characterization of these specimens. The nanoformulations include the methodology and characterization of resulting nanoparticles.

628322 Molecular Dynamics Simulation of the Sintering of Titanium/Aluminum Core/Shell Nanoparticles

Abstract:

Bimetallic core/shell nanoparticles (NPs) yield intriguing microstructures after being processed into final products. Direct metal laser sintering (DMLS) and selective laser sintering (SLS) are two powder fusion processes for metallic NPs, with the latter being the preferred additive manufacturing technique for these materials. Among bimetallic NP candidates for SLS, Ti/Al core/shell NPs have found much interest lately. However, not much is known about their sintering states during this process. To fill this knowledge gap, we performed a series of classical molecular dynamics (MD) simulations to roughly mimic the conditions of SLS for Ti/Al core/shell NPs. We further investigated the role of Ti core volume fraction (3, 12, and 30%) on the resultant uniaxial tensile properties of the sintered NPs during various sintering states. For this purpose, we created a "chain" model from five single thermally equilibrated Ti/Al core/shell NPs with weak neck connections by solid-state sintering at 298 K. The chains were then heated to 800 K with slow and fast heating rates of 0.04 K/ps and 0.2 K/ps, respectively. Next, they were relaxed at this high temperature (to yield stronger NP-to-NP neck connections), where after they were cooled back to 298 K with a cooling rate of 0.08 K/ps. They were then relaxed at this temperature for different periods of 1, 4, and 10 ns. To determine the characteristics of the heat sintering process, we measured the mean square displacement (MSD) of the atoms and squared radius of gyration from the trajectory data. MSD as a function of temperature indicates the extent of sintering and the quality of the final sintered products. We further investigated the structural evolution of the atomic configurations during the sintering process. Overall, heating at a rate of 0.04 K/ps yields lower neck-formation temperatures for all NP core volume fractions. Also, on average, an increase in the Ti core volume fraction of the NPs from 3% to 30% causes a drop in the neck-formation temperatures at both heating rates. In a follow-up procedure, we subjected the sintered NP products to uniaxial tension with strain rates of 0.001, 0.01, and 0.1 %/ps along the x-direction. For all cases of core volume fractions, the Young's moduli were determined to be within 40 GP, indicating that Ti core volume fraction has a negligible effect on the elastic tensile properties of the Ti/Al core/shell NP chain products. Our results further indicate that a larger Ti core volume fraction yields stronger chain structure and, hence, a higher tensile strength. The effect of heating rate on the tensile strength of the final products with larger core volume fractions is more pronounced. Also, high strain rates result in an improvement in the tensile strength and ductility of the final products. It is expected that insights gained from this work will promote the SLS process of bimetallic core/shell NPs.

628732 Elucidating the Molecular Mechanism of Diffusive and Electrophoretic Ion Transport Under Single Digit Nanoconfinement

Abstract:

Abstract Withheld

632172 High-Throughput Computational Analysis of the Role of Finite Temperature in the Optical Response of 2D Materials

Abstract:

Stable two-dimensional materials are promising for nanoelectronic applications due to their relatively small form factor, and their great variety of chemical and optoelectronic properties [1]. This allows for the design of light-weight and low-cost materials for optoelectronic devices for applications such as single-electron transistors [2], solar cells [3], and lasers [4]. Here, we utilize first-principles density functional theory (DFT) to investigate the optoelectronic function of these monolayer materials, including the role of electron-phonon interactions.

DFT is the standard modeling methodology for studying the electronic properties of solid-state materials. However, the common T = 0K approximation may result in an erroneous understanding of the material properties [5]. This is especially the case for low-dimensional materials with low-screening that results in increased electron-phonon interactions [6]. We present a high-throughput DFT and beyond-DFT approach to study of the effect of finite temperature on the optical properties of 2D materials, incorporating the role of phonons through a semi-classical approach [7]. We show that the strength of electron-phonon interactions is highly dependent on the bond ionicity within the crystal. We show that this framework allows for a systematic theoretical exploration of new materials for solar energy conversion.

[1] Q. Ma, G. Ren, K. Xu, and J. Z. Ou, "Tunable Optical Properties of 2D Materials and Their Applications," Advanced Optical Materials, vol. 9, no. 2, p. 2001313, 2021, doi: https://doi.org/10.1002/adom.202001313.

[2] M. Javaid, D. W. Drumm, S. P. Russo, and A. D. Greentree, "Surface-gate-defined single-electron transistor in a MoS 2 bilayer," Nanotechnology, vol. 28, no. 12, p. 125203, Mar. 2017, doi: 10.1088/1361-6528/aa5ce0.

[3] A. Pospischil, M. M. Furchi, and T. Mueller, "Solar-energy conversion and light emission in an atomic monolayer p–n diode," Nature Nanotech, vol. 9, no. 4, pp. 257–261, Apr. 2014, doi: 10.1038/nnano.2014.14.

[4] B. Zhang et al., "Recent Progress in 2D Material-Based Saturable Absorbers for All Solid-State Pulsed Bulk Lasers," Laser & Photonics Reviews, vol. 14, no. 2, p. 1900240, Feb. 2020, doi: 10.1002/lpor.201900240.

[5] M. Zacharias, C. E. Patrick, and F. Giustino, "Stochastic Approach to Phonon-Assisted Optical Absorption," Phys. Rev. Lett., vol. 115, no. 17, p. 177401, Oct. 2015, doi: 10.1103/PhysRevLett.115.177401.

[6] T. Sohier et al., "Enhanced Electron-Phonon Interaction in Multivalley Materials," Phys. Rev. X, vol. 9, no. 3, p. 031019, Aug. 2019, doi: 10.1103/PhysRevX.9.031019.

[7] M. Zacharias and F. Giustino, "Theory of the special displacement method for electronic structure calculations at finite temperature," Phys. Rev. Research, vol. 2, no. 1, p. 013357, Mar. 2020, doi: 10.1103/PhysRevResearch.2.013357.

632407 *Ab Initio* Molecular Dynamics Simulations of the Hydration of Zwitterions

Abstract:

Biofouling is a global problem caused by the gradual accumulation of biomolecules and microorganisms, such as barnacles and algae, on a surface. For example, it increases the hydrodynamic drag on vessels' sailing and causes pitting and crevice corrosion of vessels' metal surfaces. Developing a non-toxic and eco-friendly biofouling-resistant coating—which is also effective in different marine environments—is essential to minimize the biofouling effects. Zwitterionic polymers are promising candidates to this end due to their electrostatically driven strong hydration. Experimentally, it is found that the antibiofouling property of zwitterions increases as the zwitterionic separation decreases. Employing *ab initio* molecular dynamics, we investigate hydration of multiple zwitterions of varied molecular structures. We further study their strong hydration's assumed origin, the non-covalent interactions between the zwitterions and their solvent waters, using symmetry-adapted perturbation theory. Our simulations show that the zwitterionic-hydration is varied with the separation distance of oppositely charged moieties, and the induction interaction with hydrogen-bonded water plays a critical role.

632361 Green Synthesis of One-Dimensional Silver Nanostructures Using Tannic Acid Simultaneously As Reducing, Stabilizing, and Capping Agent-Parametric Study

Abstract:

The green synthesis of metallic nanostructures has risen as a promising alternative for the conventional chemical and energy-intensive methods. Most novel green methods include the use of biological reagents such as living microorganisms (e.g. bacteria, fungus, microalgae), or plant-based reagents such as vitamins, sugars, and bark/leave extracts, to name a few. Plant-based reagents are more favorable in terms of higher conversion rate of metal ions to metal atoms, non-pathogenic behavior, and process simplicity. Tannins are a particular group of plant-based polyphenols that can be used simultaneously as reducing, stabilizing, and capping agents for the synthesis of metallic nanoparticles at room temperature. Despite the few studies carried out on the synthesis of silver nanowires (AgNWs) using tannins, the effect of different reaction conditions as well as underlying mechanisms need more investigation and in-depth analysis.

In this work, we report using tannic acid, which is a hydrolysable form of tannin, for the synthesis of AgNWs at room temperature without using any external capping agent or stabilizer. We have investigated the effect of different reaction conditions including tannic acid concentration, silver precursor concentration, pH, and stirring rate. Such factors were analyzed to identify the most optimal condition for the synthesis of silver nanowires with the highest yield. Furthermore, the photosensitivity of the reduction reaction is analyzed by investigating the effect of light. In addition, a kinetic study was performed using the UV-Visible spectroscopy (UV-Vis) to analyze the reduction kinetics at different illumination conditions. The Transmission Electron Microscopy (TEM), Scanning Electron Microscopy (SEM), and Energy Dispersive X-Ray (EDX) were also used to characterize the size and morphology of AgNWs, and to compare the yield of AgNWs corresponding to different reaction conditions.

The results demonstrated that the tannic acid-mediated reduction process is highly photosensitive, and the rate of reduction (Ag^+ conversion to Ag^0) can be significantly affected by environmental light. We also report the synthesis of AgNWs with moderate to high yield at acidic conditions, controlled by the amount of light being transmitted into the system. Furthermore, the kinetic analysis sheds light on the reduction process and aims to illustrate how the reduction rate can be used as a quantitative knob to maximize the concentration of multiple twinned seeds that are favorable for the one-dimensional (1D) nanostructures growth including AgNWs. The synthesized AgNWs can be used for the preparation of electrically conductive materials such as transparent conductive films (TCFs) for flexible electronics.

632669 Peptide-Mediated Fabrication of Pd–Au Nanocatalysts: Effect of Metal Composition and Catalyst Design Approach on Nitrite Reduction in Water

Abstract:

Here we examine the use of materials binding Pd4 peptide, originally isolated with Pd affinity with ability to bind gold, for the synthesis of catalytically active palladium decorated gold (Pd-on-Au) and palladium-gold (PdAu) nanoparticles (NPs). Since the preparation of Pd-on-Au and PdAu NPs requires variety of chemical, harsh conditions, and prolong circumstances, a water-based one-pot fabrication route at room temperature using Pd4 peptide was adapted for rational design of catalytic NPs for practical use. The prepared NPs were then exploited for catalytic degradation of nitrite (NO_2^-) as a crucial contaminant presenting in water resources. The developed peptide-mediated catalysts showed 3.6 times increase in NO_2^- reduction comparing to monometallic Pd NPs in comparable TOF with previous reports. The prepared NPs showed high selectivity toward nitrogen gas production over ammonia.

632656 Colloidal Synthesis of Zinc Tin Phosphide Nanocrystals

Abstract:

Our growing energy usage has created an urgent need for sustainable sources of electricity and optoelectronic devices including photovoltaics and lighting. Successful development of such technologies relies on the discovery of scalable photonic materials that can compete with current state-of-the-art III-V (GaN, GaAs) semiconductors but have low toxicity and processing costs. Composed of abundant, non-toxic, and recyclable elements, we present the II-IV-V_2 system $ZnSnP_2$ as a candidate for such applications. Ternary pnictides such as $ZnSnP_2$ offer the potential of tuning the band gap via cation disorder at fixed compositions. Realizing the potential of $ZnSnP_2$, however, has been difficult to accomplish. Syntheses have typically led to impure materials due to the difficulty of controlling the presence of defect phases and anti-site defects, leading to uncontrollable optoelectronic properties. In this work, we present a hot-injection colloidal synthesis method as a defect-limited route to $ZnSnP_2$. We obtain nanocrystals approximately 4 nm in size; we determine crystal phase via x-ray diffraction measurements, pair distribution function analysis, and transmission electron microscopy imaging and further study the local structure of our nanocrystals via x-ray absorption spectroscopy. Optical spectroscopy measurements taken over the growth period of the nanocrystals provide insight on the evolution of structure-property relationships in the synthesized nanocrystals.

622844 Synthesis of Water-Dispersible $Ti_3C_2T_z$ Mxene Nanosheets By Molten Salt Etching

Abstract:

This study uses a molten salt (SnF_2) etching method to synthesize water-dispersible $Ti_3C_2T_z$ (MXene) from the parent Ti_3AlC_2 MAX phase. The new approach avoids the use of hazardous HF for etching. The etched $Ti_3C_2T_z$ clay was washed with KOH, intercalated with DMSO, washed with water, bath sonicated, and centrifuged. In this processing method resulted in a stable $Ti_3C_2T_z$ nanosheet dispersion in water with a ζ potential of -31.7 mV. The treatment of $Ti_3C_2T_z$ with KOH resulted in stable dispersion due to the addition of -OH terminal groups to $Ti_3C_2T_z$. X-ray diffraction and Electron microscopy confirms the formation of $Ti_3C_2T_z$ clay with substantial d spacing when compared to the traditional HF-etched MXene. This work is the first to use molten salt etching to prepare colloidally stable aqueous dispersions of $Ti_3C_2T_z$ nanosheets successfully. The $Ti_3C_2T_z$ nanosheet film and clay produced by this method also show excellent specific and areal capacitance.

629457 Plasmonic Photocatalysis for Gas-Phase Degradation of Total Volatile Organic Compounds: Theory, Experimentation, and Catalyst Stability

Abstract:

Plasmonic nanomaterials have become a strong contender for improving the efficiency of photocatalytic degradation of air pollutants. This study demonstrates an easy and scalable fabrication method, employing electron beam evaporation and rapid thermal annealing, for producing plasmonic photocatalysts. Samples were made by either coating silver on, or layering silica-protected (i.e., silica-coated) or unprotected (i.e., uncoated) silver beneath, the photocatalyst (either zinc oxide or titanium dioxide). Stability and catalytic performance for gas-phase toluene degradation was assessed by monitoring total volatile organic compound (TVOC) concentration versus ultraviolet-A (UV-A) illumination time in a recirculating batch reactor (plate-type flow-through). Samples were characterized using a variety of spectroscopic methods, electron microscopy, and X-ray diffraction. Finite difference time domain (FDTD) simulations demonstrate that the UV-A light-induced electric fields around silica-protected and unprotected silver nanoparticles extend significantly beyond the nanoparticle surfaces, allowing contact between the fields and the photocatalyst, and justifying the catalyst design. Experimental results, and analysis of reaction kinetics, show that silver-coated photocatalysts and layered photocatalysts with unprotected silver exhibit initially high TVOC degradation rates, but suffer from deactivation, attributed to silver oxidation. Layered photocatalysts with silica-protected silver exhibited improved stability versus unprotected samples and improved performance over titanium dioxide. The results of this study demonstrate that a silica layer can help slow down fouling of silica nanoparticles, and that silica-coated silver nanoparticles not only help speed up reaction rate overall but also appear to speed up release of adventitious hydrocarbons from the photocatalyst surface during the initial phase of a reaction. The layered fabrication approach developed and employed in this study for application to gas-phase photocatalysis may enable plasmonic photocatalysis by offering a simple, scalable fabrication method more reliable than the more prevalent colloidal synthesis.

632613 Chiral Self-Assembled Structures of Bowtie Shape: Synthesis, Optical Properties, and Deep Learning-Based Modeling.

Abstract:

Chiral nanostructures offer new solutions in many technological fields, including biosensing, chiral catalysis, enantioselective separation, and quantum computing. While many factors make chiral inorganic nanostructures significant to these areas, the unique characteristic that drives research in this area is their ability to exhibit high chiroptical activity. To date, many researchers are focused on materials combining nano-, meso-, and micron-scale structural elements since such hierarchical structures can rotate light polarization in a wide spectral range. However, top-down approaches, such as ion-beam lithography, extensively used for the growth of such helices, are limited for the fabrication of nanoscale elements and have a low mass production efficiency.

Here we developed a facile bottom-up approach for the formation of microscale helices of bowties shape. Depending on the enantiomer of chiral molecules used bowties have a right- or left-handed twist with 100% enantiomeric excess. Transmission electron microscopy of bowties allowed to establish their assembly mechanism and growth. Bowties are formed by self-assembly of nanoparticles ~ 3.4 nm diameter that are arranged into stacked twisted nanosheets (**Figure 1a**). The latter have a thickness of about 100 nm and a 50 nm separation between each other. Careful adjustment of synthesis conditions allows growing bowties with desired length, width, thickness, and twist, which, in turn, lead to the strong chiroptical activity that could be carefully tuned in a wide spectral range. Computations of bowties chiroptical activity has shown that the circular dichroism and g-factor spectra experience a blue-shifting tendency as the pitch of bowties decreases (**Figure 1b**). To predict bowties optical properties, we also developed an artificial neural network that prognosticates *g*-factor values and spectra peak positions with reasonable accuracy (**Figure 1c**).

In summary, we report here the self-assembly of bowties consisting of twisted stacked nanosheets. The morphological properties of bowties are tuned by varying the chemical conditions tuning their optical properties. We believe that our results open new horizons for designing chiroptical materials and metamaterials with a strong and tunable polarization rotation essential for multiple emerging technologies.

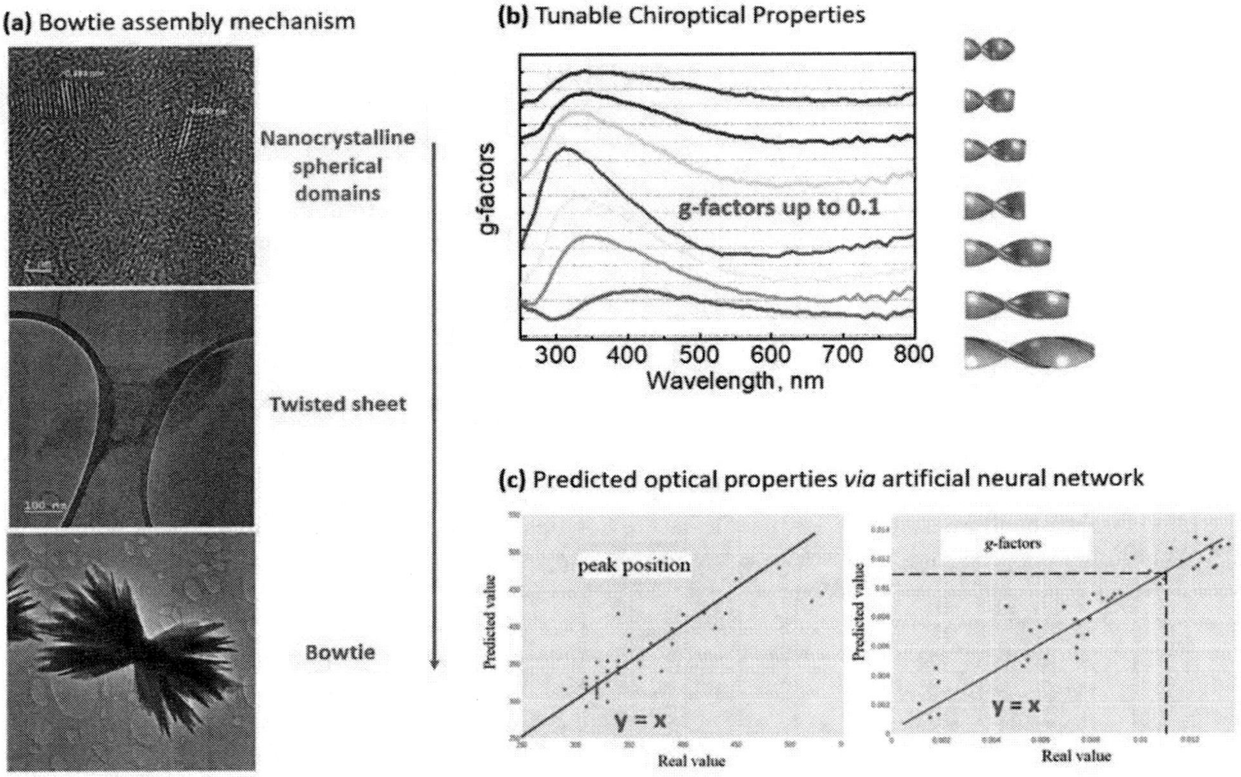

Figure 1. (a) Transmission electron microscopy images of bowtie and its intermediate stages; (b) *g*-factor spectra of bowties of different pitch; (c) peak positions of *g*-factor spectra and g-factor values predicted using artificial neural network.

628094 MOF-Derived Metal Oxide Composites for High Performance Energy Storage

Abstract:

Metal-organic frameworks (MOFs) are crystalline porous structures composed of metal ion centers and organic linkers coupled through coordination bonding. The intrinsic high specific surface area, controllable pore size and good thermal stability of MOF structures makes them good candidates to be used as porous templates. Through a novel coordination modulation method, it is possible to control the size, morphology, and structural properties of MOF structures by incorporating secondary ligands to modulate the coordination equilibrium with the same chemical functionality as the primary organic linkers to act as capping agents and to adjust the coordination interactions between metal ions and organic linkers and thus resulting modulated MOFs in terms of morphology and crystal size. This paves the way to develop and synthesize MOF structures with favorable characteristics to be used as sacrificial templates to synthesize interconnected porosity and high surface area.

The advanced approaches to develop high performance electroactive materials mainly focus on introducing more electrochemically active sites, shortening ionic pathways, faster charge transfer kinetics. This will be only achieved by careful selection of appropriate electroactive material as well as rational design of hybrid electrode architecture using conductive scaffolds such as graphene. We tend to focus more on exploiting methods to improve the electrochemical performance of metal oxides composites by trying to understand better the metal oxide composite formation and the metal oxide/graphene interfacial interactions and its effect the electrochemical kinetics of our composite structure.

In this work, we adopt a rational strategy to synthesize $NiCo_2O_4/Co_3O_4$ metal nanostructures. Highly porous ZIF-67 nanocrystals are firstly produced through coordination modulation process. By carefully etching these structures, ZIF-67/Ni-Co layered double-hydroxide template is formed. The as-prepared templates are subsequently calcinated to obtain the $NiCo_2O_4/Co_3O_4$ metal nanostructures. Later, to prevent the agglomeration and improve the charge transfer properties of our structures, the surface of as-prepared nanostructures is functionalized using a silanization process and mixed with graphene oxide dispersion to form self-assembled $NiCo_2O_4/Co_3O_4$-rGO composite structures. We report $NiCo_2O_4/Co_3O_4$-rGO composite structure with a capacity of 969 F g^{-1} at current density of 1 A g^{-1} showing a remarkable stability of showing 93% of initial capacitance after 10000 cycles of charge-discharge using electrodes with a high mass loading (~4 mg cm^{-2}).

628277 Nanostructured Zinc Anodes for Ultra-Safe High-Energy Rechargeable Batteries

Abstract:

Metallic zinc as a rechargeable anode material for aqueous batteries has gained tremendous attention with merits of intrinsic safety, low cost, and high theoretical volumetric capacity (5,854 mAh/cm3). Among zinc-based batteries, Zn-air batteries are promising with the highest theoretical volumetric energy density (4,931 Wh/L). Rechargeable zinc anode has recently achieved progress in neutral electrolytes, yet developed slowly in alkaline electrolytes, which are kinetically favorable for air cathodes. Passivation, dissolution, and hydrogen evolution are three main reasons for irreversibility of zinc anodes in alkaline electrolytes, which limits the rechargeability and usable energy density. In this talk, I will present our recent works on using nanoscale material design to overcome passivation, dissolution, and hydrogen evolution issues of zinc anode, towards a deeply rechargeable zinc-based battery. I will also introduce the battery-gas chromatography quantitative analysis, as well as in situ microscopy methodologies we have developed, to quantify gas evolution side reaction, as well as visualize the reaction on electrodes during operation.

630784 Tailoring the Assembly of Cost-Effective Milled Silicon and Exfoliated Graphene for High Capacity, High Rate-Capable Lithium-Ion Battery Anodes

Abstract:

Silicon as anode material has been very promising to satisfy the high energy density demands for electric vehicle (EV) applications due its extraordinary theoretical capacity. However, silicon undergoes huge volume expansion(~ 300%) during lithium-ion storage. One approach to address this issue is nanostructuring of silicon(30-50nm) to minimize the stress generated during volume expansion. Although, nanostructuring has resolved most of the issues associated with silicon anodes, these nanostructure morphologies involve multiple and expensive production steps which makes it impractical on an industrial scale. One attractive option to reduce the cost, substantially, is by using milled low-cost metallurgical Si(200-300nm) which is often the recycled from the Si scrap that is easily obtained from solar or semi-conductor industry. To accommodate the mechanical stress caused due to the volume expansion of milled silicon, graphene or reduced graphene oxide is frequently used. Graphene not only acts as a buffer against the volume expansion to preserve structural integrity of the electrode, but also increases the overall conductivity of the electrode. Most of the Graphene available commercially is made from reduction of graphene oxide by heat, chemical or plasma treatment leading to increase in cost in the final product.

In our study we are using exfoliated graphene which is produced in-house from a batch Taylor-Couette reactor set-up. The Taylor-Couette setup used in this study included the rotation of the outer cylinder while the inner cylinder is still. Graphite flakes are dispersed in an aqueous solution with a dispersant and a stabilizer followed by introducing the solution into the gap in the Taylor-Couette system. Substituting our mechanically exfoliated graphene instead of commercial reduced graphene oxide in milled silicon anodes not only greatly reduces the cost by several orders of magnitude, due to direct synthesis of graphene from graphite as opposed to reducing GO, but also makes the overall process more facile.

Several graphite pre-cursors for the Taylor-Couette system were studied which resulted in mechanically exfoliated graphene sheets with different lateral dimensions and varied level of defects. The interplay of graphite pre-cursor and electrochemical performance was probed. Our results show that exfoliated graphene can provide high capacities and stable retention over several cycles and successfully replace commercially obtained rGO. More importantly, the resultant electrodes are able to provide impressive rate capabilities of 800-1000 mAh/g at 2C. Exfoliated Taylor-Couette Graphene not only shows great potential for cost reduction but also enhanced electrochemical performance.

631034 *Layer-on-Layer High Sulfur-Loading Cathodes for Lithium Sulfur Batteries Via Air-Controlled Electrospray*

Abstract:

Among the diverse next-generation batteries, lithium-sulfur (Li-S) batteries have been recognized as one of the most promising choices beyond lithium-ion batteries (LIB) especially in the field of electric vehicles, because of its low cost, and its high theoretical specific energy of 2510 Wh/kg, 10 times of the LIB. However, Li-S batteries are still facing a few challenges, including large volume expansion, poor conductivity of end products, sluggish kinetics of polysulfides, and the "shuttle effect", etc. The shuttle effect is caused by the dissolved polysulfides spontaneously diffusing to anode, where it will be immediately reduced by lithium to form the insoluble and insulating short-chained polysulfides, Li_2S_2 and Li_2S, passivating the anode and losing active materials. Another implicit drawback of the state-of-art Li-S battery is the low sulfur loading around 1mg/cm^2 which limits the areal capacity to be around or less than 1mAh/cm^2. Increasing sulfur loading has been identified as a crucial future direction in recent review articles, and definitely a critical step toward commercialization of Li-S batteries. Nevertheless, shuttle effect is often more severe in high-loading cathodes due to the higher diffusion gradient of polysulfide concentration.

To address these challenges, we employed a "layer-on-layer" cathode structure with sulfur-infiltrated porous carbon layer alternating with thin reduced graphene oxide (rGO) layers, which serve as a "fishnet" to entrap polysulfide from diffusion toward anode. There are other polysulfide adsorbents reported in the literature, but many of them are insulative rare metal oxides, while our choice of rGO has the advantage of high conductivity, low cost, and vast availability as a common industrial material. The layer-on-layer structure can be facilely fabricated by air-controlled electrospray (ACES), which keeps the active material dispersion as tiny droplets by air flow and electric field, and thus allows rapid drying with less cracking and uniform deposition. The cathodes were analyzed by various characterization techniques, including SEM, TGA, EIS, etc, to unveil the layer-on-layer structure, the improved conductivity, and the reduced shuttle effect. Our cathodes target at 5mg/cm^2 sulfur while demonstrating superior specific capacity and charge-discharge cycling stability, compared to 2mg/cm^2 maximum by the conventional slurry casting.

631326 Synergy of Graphene Nanoribbons and Graphene Sheets for High-Rate Lithium-Sulfur Batteries

Abstract:

According to the increasing demands for shortening the battery charging time, high current rate (C-rate) performances become more significant in practical applications. With a higher theoretical capacity, lithium-sulfur batteries are treated as promising candidates for the next-generation batteries. In this work, the utilization of graphene nanoribbons (GNRs) exhibits the benefits in conductivity and other electrochemical performances, especially for high-rate applications. With air-controlled electrospray as the method, carbon encapsulated sulfur particles, poly(acrylic acid), reduced graphene oxide (rGO) sheets, and GNRs are mixed and directly deposited onto the carbon coated aluminum collector, to employ as the cathode. The scanning electron microscopy (SEM) imaging exhibits that the two-dimensional structure of GNRs helps construct inter-connected networks. This improved structure of cathode can increase the electroconductivity, confirmed by the electrochemical impedance spectroscopy (EIS), and modify the porosity, indicated through pore size distribution profiles. In this way, the polysulfides can be more efficiently trapped and utilized, realizing the promising behavior with faster charge transfer. In terms of the cycling performance at 0.2 C, the batteries with GNRs can perform 18% higher in capacity than those without GNRs, without decreasing the charge retention. According to the rate-capability tests, systems with GNRs can achieve enhanced performance compared to batteries with precursor carbon nanotubes (CNTs), especially at high C-rates. At 2 C, with 80 wt % of graphene-based materials as GNRs, batteries can achieve an increase in capacity by 78% and 41% compared with systems without GNRs and those with CNTs, respectively. Accordingly, the results testify the synergy of GNRs and rGO sheets in Li-S batteries.

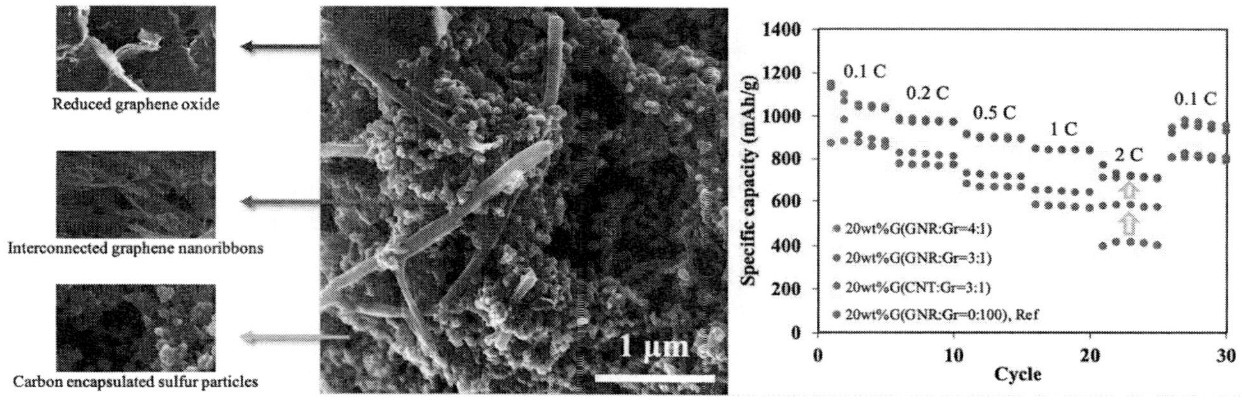

629657 Biomineralization of Hydrogen Evolution Photocatalysts

Abstract:

While photocatalysts offer a promising route to efficient and clean hydrogen production, the multi-step, high temperature, solvent based syntheses typically utilized to prepare these photocatalysts can limit their scalability and sustainability. Bio-synthetic routes for photocatalyst production—occurring at room temperature and in aqueous conditions—address these problems, but the as-synthesized materials are often not high-performing due to lack of control over product size and inability to form composites and heterojunctions. Herein, we demonstrate that composite photocatalysts synthesized through bio-inspired room temperature, aqueous routes can achieve hydrogen evolution rates comparable to those made by traditional means. We have investigated the enzyme-based biomineralization of photocatalysts and supports as well as the assembly of these materials to form high-performing composite photocatalysts. We have isolated an enzyme, cystathionine γ-lyase, that turns over L-cysteine to HS^-, which further reacts with metal ions in solution to form metal sulfide nanocrystals. Through this biomineralization route, we have demonstrated the size-controlled synthesis of CdS nanocrystals that are active for photocatalytic hydrogen evolution. Additionally, the biomineralization approach reduces graphene oxide to produce reduced graphene oxide (rGO), a conductive support for photocatalysts. Ligand exchange, involving replacement of the negatively charged native L-cysteine ligands on CdS for positively charged cysteamine ligands, mediates self-assembly of CdS nanocrystals on rGO in solution at room temperature. The visible light hydrogen evolution rate of the resulting CdS/rGO photocatalyst was measured at 3300 μmol hr^{-1} g^{-1}, which is nearly double the rate of the unsupported CdS. This represents, to our knowledge, one of the highest reported rates for a CdS/rGO nanocomposite photocatalyst. Furthermore, the biomineralization of a heterojunction CdS/ZnS photocatalyst was also investigated by Zn addition through post-synthetic addition of zinc acetate to biomineralized CdS nanocrystals. Adding zinc acetate to CdS nanocrystals was shown to controllably increase the radiative recombination rate of the nanocrystals and to, thereby, improve the photocatalytic hydrogen evolution rate from 1300 μmol hr^{-1} g^{-1} to 3200 μmol hr^{-1} g^{-1}.

632213 Multifunctional Hollow Carbon Nanorods with Open Ends for Energy and Biomedical Applications

Abstract:

Developing multifunctional materials for various applications is a fundamentally challenging task. In particular, hollow structured carbon materials have been attracting much attention in the past two decades. Multifunctional hollow carbon structures can offer integrated properties for energy, environmental and biomedical applications. However, it is still an intellectually challenging task and practically difficult to prepare multifunctional hollow nanomaterials with unique morphologies and desired integrated properties. Herein, we report novel hollow carbon nanorods with open-ended structures for highly stable lithium metal anode and potential biomedical applications. The novel hollow carbon nanobeads structures were fully characterized by SEM, TEM, XRD, and Raman analysis. In this presentation, we will highlight our latest efforts on the preparation, characterization of the unique hollow carbon nanorods with open ends, and demonstrate their promising performances in energy and biomedical applications.

630686 Aerosol Nanocomposite Systems Comprised of Cell Membrane-Coated Nanoparticles for the Treatment of Pulmonary Diseases

Abstract:

The lungs are an attractive route for drug delivery applications that offer many advantages, such as a large surface area, high blood flow, and limited enzymatic activity of the lungs, in addition to the ability to directly target lung tissue through aerosol delivery of therapeutics. There has been increasing interest in the aerosol administration of nanoparticles (NP) to achieve controlled delivery of drugs via the lungs to allow for extended release of therapeutics (in particular those that are poorly water soluble), direct targeting to the site of disease, and the potential for NP to penetrate physiological barriers in the lungs such as mucus, tumors, and the air-blood barrier. Our group has developed nanocomposite microparticles (nCmP) that involve drug-loaded biodegradable NP that are encapsulated into microparticles via spray drying. Micron-sized nCmP have the ability to be effectively delivered to the alveolar region of the lungs, while the NP that disassociate from the nCmP have the ability to avoid phagocytosis by alveolar macrophages, thereby acting as a 'Trojan horse' system.

Cell membrane coated nanoparticles (CMCNP) have attracted recent attention in the field of NP drug delivery owing to the ability to target diseased tissue and be used in a wide variety of medical applications such as imaging and detoxification. CMCNP are comprised of a base NP core (biodegradable or otherwise) that is coated with a cell membrane comprised from cell such as platelets, red blood cells, immune cells, and endothelial cells. In this work we have developed CMCNP comprised of a drug-loaded biodegradable acetalated dextran core coated with two different lung epithelial cell lines (A549 and H441). The CMCNP were compared with two lipid-coated NP systems (anionic DPPC and cationic DOTAP) and two polymer-coated NP systems (polyethylene glycol (PEG) and polyvinyl alcohol (PVA)). The NP were encapsulated with fluorescent curcumin (CUR) as a model poorly water soluble therapeutic. The CMCNP were then spray dried to form aerosol nCmP. Overall, we sought to determine how varying surface properties affected the ability of NP to internalize into and/or transport across an *in vitro* pulmonary epithelial barrier in addition their impact on size, surface charge, physicochemical properties, and aerosol dispersion properties.

The NP systems were all ~200nm in diameter and exhibited appropriate surface charges according to their respective coatings - negative for A549 NP, H441 NP, and DPPC NP; positive for DOTAP NP, and neutral for PEG NP and PVA NP. No significant NP toxicity was observed when H441 cells were exposed to the NP, and controlled CUR release from the NP was achieved. CMCNP more readily internalized and transported across a pulmonary H441 cell monolayer and nCmP were produced the exhibit favorable aerosol dispersion properties. Furthermore, A549 CMCNP and PVA NP were administered intratracheally to mice to demonstrate the ability of the NP to be delivered to the alveolar region of the animals' lungs. In summary, this research demonstrates the first time epithelial lung-based CMCNP and nCmP were fabricated. These systems have the potential to be used in pulmonary drug delivery applications for the treatment of a variety of pulmonary diseases. Future studies will include the elucidation of the ability of CMCNP to cross the air-blood barrier for systemic NP delivery or NP homing to disease tissue.

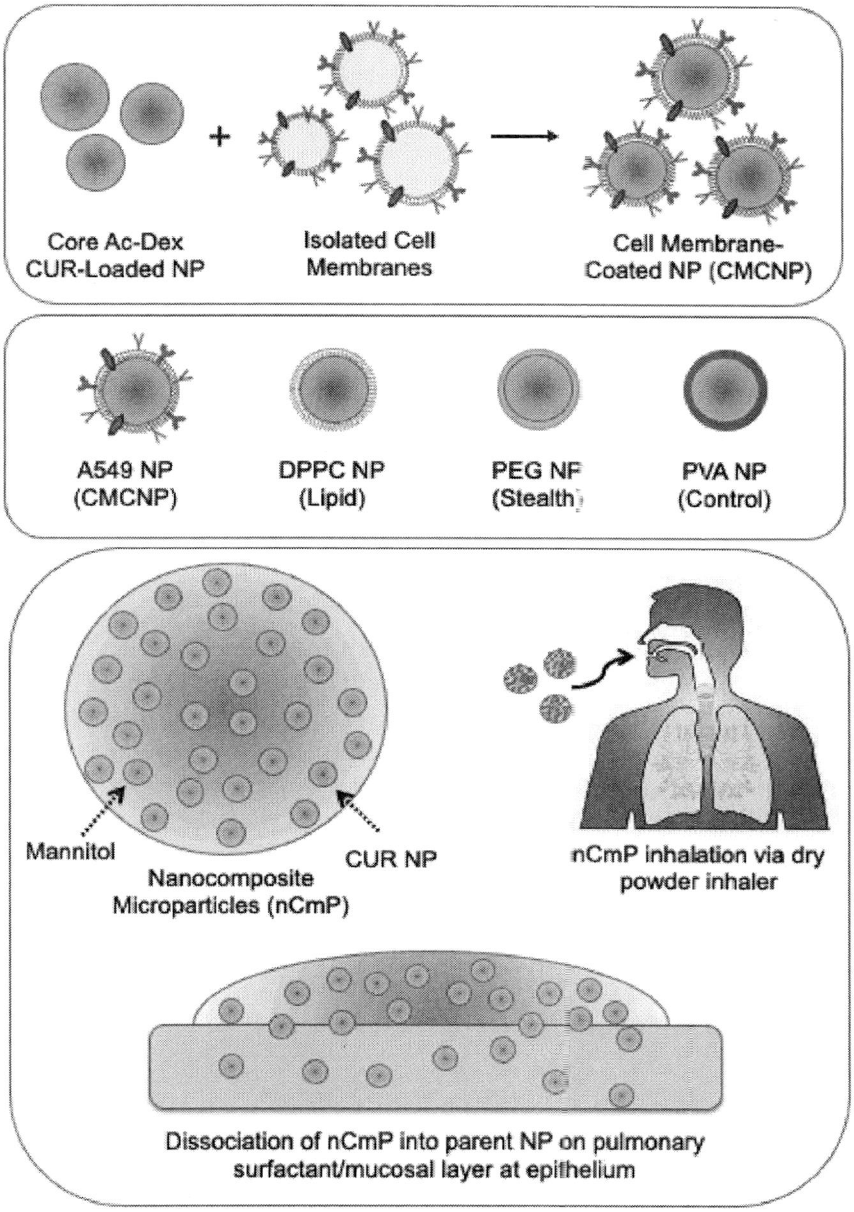
Core Ac-Dex
CUR-Loaded NP
Isolated Cell
Membranes
Cell Membrane-
Coated NP (CMCNP)
A549 NP
(CMCNP)
DPPC NP
(Lipid)
PEG NP
(Stealth)
PVA NP
(Control)
Mannitol
CUR NP
Nanocomposite
Microparticles (nCmP)
nCmP inhalation via dry
powder inhaler
Dissociation of nCmP into parent NP on pulmonary
surfactant/mucosal layer at epithelium

631908 Effect of Drug Hydrophobicity on X-Ray-Triggered Drug Release from Peg-PLA/$CaWO_4$ Nanoparticles: A Study of Stereoisomers of paclitaxel

Abstract:

For many locally advanced tumors, the chemotherapy-radiotherapy (CT-RT) combination ("chemoradiation") is currently the standard of care. Intratumoral (IT) CT-based chemoradiation has the potential to overcome the limitations of conventional systemic CT-RT (side effects). For maximizing the benefits of IT CT-RT, our laboratory has developed a radiation-controlled drug release formulation, in which anticancer drug paclitaxel (PTX) and radioluminescent $CaWO_4$ (CWO) nanoparticles are co-encapsulated with poly(ethylene glycol)-poly(lactic acid) (PEG-PLA) block copolymers ("PEG-PLA/CWO/PTX"). These PEG-PLA/CWO/PTX nanoparticles enable radiation-controlled release of PTX, and are capable of producing sustained therapeutic effects lasting for at least one month following a single IT injection.

This talk will focus on discussing our recent finding about the effect of the stereochemical structure of PTX on the efficacy of this PEG-PLA/CWO/PTX formulation. Stereochemical differences in two different PTX compounds were characterized by 2D heteronuclear NMR and circular dichroism measurements. The difference in PTX stereochemistry was found to significantly influence the efficacy of the formulation, both in vitro and in vivo. The two compounds showed similar performances when used as free drugs. However, the subtle stereochemical difference significantly influenced their X-ray-triggered release kinetics from the PEG-PLA/CWO/PTX nanoparticles. This difference was manifested in the intratumoral pharmacokinetics and eventually in the survival percentages of test animals (mice) in a human tumor xenograft study. This study demonstrates the importance of the stereochemistry of a drug in a therapy based on a controlled release formulation.

632225 Liposome and Polyelectrolyte Layers Derived Single Shot Vaccine Platform for Controlled Release of Inactivated Chikungunya Virus

Abstract:

Immunization through inactivated virus enhances the safety profile of the vaccine and is one of the most efficient technique to protect against infectious diseases. However, inactivated vaccines require multiple rounds of vaccinations that often lead to missing or mistimed doses thereby limiting their potency. This multidose requirement further limits mass immunization during the outbreak of an infectious disease such as chikungunya virus (CHIKV) in a place without herd immunity. Single shot vaccines capable of releasing the antigen intermittently can potentially be employed to overcome the shortcomings of multi-dose vaccines and provide herd immunity in shorter period of time. A sustained release of antigen can help in maintaining high concentration of antibodies in the system and programmed delayed release can induce strong booster effect. In this study, we engineered a single shot vaccine delivery platform comprising liposomes (LNP) embedded polyelectrolyte multilayer (PEM) lattice to provide controlled spatiotemporal release of CHIKV antigen and thus induce long term immunity. The CHIKV virus inactivated using γ-radiations in presence of MDP complex (MDP-iCHIKV) was injected in mice and tested for production of antibodies. The MDP-iCHIKV were then encapsulated in LNP followed by embedment in PEM films so as to obtain a programmed release of antigen at different time points. We performed the characterization of LNP and LNP-PEM assembly using dynamic light scattering method to determine the size and zeta potential measurements to determine the surface charge. Further confocal imaging and scanning electron microscopy was conducted to analyze the structure properties of the LNP-PEM assembly. Analysis of encapsulation and released CHIKV antigen concentration was performed using ELISA. In-vivo studies were performed by injecting the LNP-PEM assembly in mice followed by quantification of anti-CHIKV IgG in serum at different time points. The studies showed that MDP complex protected CHIKV viral envelop proteins while destruction of its genome by γ-radiation and also induced anti-CHIKV IgG response upon injection in mice. Significant changes in size, charge and structural properties were observed in the process of LNP embedment into PEM thus confirming the intended design of the vaccine plaform . We obtained about 75% encapsulation efficiency of CHIKV antigen in LNP and nearly 80% adhesion efficiency of LNP on PEM. In-vivo studies showed patterned increase and decrease in production of anti-CHIKV IgG indicating towards intermittent release of antigen from the LNP-PEM assembly. We are further working on optimizing the design of this single-shot vaccine platform to achieve a programmed release of antigen at multiple stages. We believe that this design for single shot vaccine platform would not only eliminate the need for multiple booster shots for inactivated vaccines but can also be potentially employed for long term immunization against a variety of pathogens.

632063 Polyethylene Glycol Camouflaged *Lumbricus Terrestris* mega-Hemoglobin for Diverse Oxygen Therapeutic Applications

Abstract:

Motivation: The newest generation of hemoglobin-based oxygen carriers (HBOCs) has found generic applications as oxygen therapeutics apart from their traditional use as resuscitative fluids. Recently, HBOCs have been used in organ and graft preservation, *ex-vivo* machine perfusion, cell growth media supplements, hypoxic tumor models, and wound healing. These diverse applications have necessitated the development of oxygen therapeutics that can be used for multiple applications. Steering away from traditional mammalian hemoglobins, this work focuses on the mega-hemoglobin (LtEc) of the annelid *Lumbricus terrestris* for use as an oxygen therapeutic. Its large size (~30 nm) reduces the possibility of *in vivo* extravasation and scavenging of endothelial-derived nitric oxide (NO). The large molecular diameter of LtEc along with its' high structural stability, low rate of auto-oxidation, moderate colloid osmotic pressure (COP), and human blood-like oxygen binding properties make it an ideal candidate for use as a resuscitative fluid in trauma treatment. Subsequently, LtEc is also a promising candidate for use in graft preservation and as an oxygen supplement in cell culture media because of its ability to stay in the reduced ferrous form for long periods of time. However, to mask the exterior protein surface and reduce potential immunogenic activity, surface conjugation is necessary. Polyethylene glycol (PEG) is a widely used and FDA approved polymer used to mask the surface of materials and improve circulatory life. We devised methods to obtain ultra-pure LtEc and surface camouflage it with PEG using thiol-maleimide chemistry. Comprehensive *in vitro* characterization and preliminary *in vivo* studies show promise for further investigation of PEG-LtEc as an oxygen therapeutic.

Methods: Purification of LtEc from *Lumbricus terrestris* worms was performed by modifying previously devised protocols from our lab with a focus on improving the purity of the resulting product. In short, Canadian nightcrawlers (*Lumbricus terrestris*) were rinsed to remove dirt and mucus prior to homogenization in a modified Tris buffer. The homogenate was centrifuged via multiple cycles to eliminate worm debris and dirt and obtain a cloudy red supernatant. Post vacuum filtration, multi-step tangential flow filtration through 0.65 μm and 0.22 μm hollow fiber filters was performed to eliminate large aggregates and bacterial components. Continuous volume diafiltrations were carried out over a 500 kDa filter to eliminate a majority of the unwanted impurities. Retentate and permeate samples were analyzed using analytical size exclusion chromatography (SEC) to track loss of proteins and the purification was concluded when the retentate reached >99% purity. *In vitro* biophysical characterization included dynamic light scattering (DLS), oxygen equilibrium analysis, rapid deoxygenation kinetics, gel electrophoresis (SDS-PAGE), and auto-oxidation kinetics. Surface modification of purified LtEc was carried out using thiol-maleimide chemistry by converting surface lysine residues to thiols and attaching PEG chains with monofunctional ends. Unreacted components were removed, and the PEG-conjugated product was analyzed using amino acid residue assays, DLS, SEC, and SDS-PAGE to confirm PEG surface modification. Oxygen equilibrium properties and other *in vitro* properties were tested and compared to unmodified material. The degree of surface camouflage was engineered by exploiting the numerous lysine residues on the LtEc surface and desirable *in vitro* properties were obtained.

Results: The modified LtEc purification protocol was successful in obtaining an LtEc product with

>99% purity, which was verified using SEC and SDS-PAGE analyses. This was an important result as the removal of non-mammalian proteins and impurities is crucial for a pharmaceutical product to be used in transfusion medicine. Biophysical characterization was performed to determine advantageous biophysical properties of LtEc such as its' large hydrodynamic diameter, moderate oxygen affinity, high oxygen cooperativity, low auto-oxidation rate, and fast deoxygenation kinetics. *In vivo* analysis in a golden Syrian hamster model corroborated that the ultra-pure product performed better in animals than product purified using older protocols with significantly better flow in arterioles and higher functional capillary density. Various degrees of surface modification with PEG resulted in PEG-LtEc products with varying biophysical properties. Even though the hydrodynamic size varied little, we observed changes in molecular masses of subunits using the SEC, SDS-PAGE, and matrix-assisted laser desorption ionization (MALDI) mass spectral analyses, further confirming successful surface camouflage. The attachment of PEG resulted in a product with higher oxygen affinity, lower oxygen cooperativity, higher auto-oxidation rates, and slower kinetic rate of oxygen offloading than LtEc. In spite of the changes to the biophysical properties, it is expected that PEGylated products will yield higher circulatory times and lesser immunogenic responses, which will be tested next.

Takeaways: The streamlined purification protocol devised in the study allows for the production of an extracellular hemoglobin with >99% purity. This purity is essential to enable a wide variety of applications of LtEc as an oxygen therapeutic. We show varying extents of camouflage of the LtEc surface with PEG and their effect on biophysical properties of the product. The product is expected to have longer circulation time *in vivo* owing to its surface masking. Future animal tests will test this hypothesis. However, the use of LtEc can be expanded to cell media supplement and graft preservation, making it a diverse oxygen therapeutic.

626632 Separations and Measurements of Single-Wall Carbon Nanotube Enantiomers

Abstract:

Aqueous two-phase extraction (ATPE) is a widely used technique for separating sub-populations of specific (n,m) structures of single-wall carbon nanotubes (SWCNTs) from synthetic mixtures through scalable, liquid-phase, processing. Most commonly, the partitioning of different (n,m) SWCNT structures is controlled by the competition of two or more surfactants for the nanotube surface. Since at least one of the surfactants is typically a chiral molecule, this competition for the nanotube surface, and the resulting partitioning of a SWCNT species has long been understood to be SWCNT (n,m) handedness dependent. However, the exact interplay and effects of different handed surfactants and validation of the surface competition model has not been directly demonstrated. Using a recently developed method for rapidly determining the surfactant concentrations necessary for extraction for specific (n,m)/(m,n) species of SWCNTs using NIR fluorescence spectroscopy, along with analytical ultracentrifugation (AUC) data on separated populations, I will present data characterizing specific (n,m) handedness isolation of multiple nanotube structures. The interpretation of these results provides values for isolating single handed (n,m) structures, and validates the surfactant surface competition model for surfactant-ATPE.

624989 A Spin-Coated Hydrogel Platform Enables Accurate Investigation of Immobilized Individual Single-Walled Carbon Nanotubes

Abstract:

Semiconducting single-walled carbon nanotubes (SWCNTs) have recently been utilized extensively as near-infrared sensors (NIR) due to their unique optical properties. Commonly, SWCNTs have been employed to detect target analytes in the solution-phase via modulations in emission intensity and/or wavelength. It is desirable to lower the limit of detection for these sensors by attempting single molecule detection using isolated SWCNTs adhered to a surface. Conventional methods of SWCNT immobilization, such as adherence to rough glass surfaces adversely affect SWCNT fluorescence. Here, we have designed a novel spin coated hydrogel platform using thin layers of alginate that mitigates SWCNT interactions with the surrounding environment, recapitulating fluorescence properties from similar solution-based sensors. The hydrogel platform is also remarkably stable, allowing for accurate spectral measurements of individual SWCNTs before and after addition of various analytes. DNA-functionalized (6,5)-SWCNTs deposited on a glass coverslip displayed an average wavelength red-shift of 3.4 nm when compared to the solution-phase, while in contrast the hydrogel platform statistically matched emission wavelength values. By looking at the width of emission wavelength histograms, the heterogeneity of SWCNTs on glass coverslips was determined to be greater by a factor of 3 when compared to SWCNTs in the hydrogel platform. Upon addition of a model analyte in calcium chloride, the DNA-SWCNTs in the hydrogel platform showed a smaller wavelength red-shift and almost no aggregation when compared to the solution-phase. Using SWCNTs with different functionalizations (DNA or surfactant) and longer than the diffraction limit of light, we show that a glass coverslip can induce spectral heterogeneity along the length of a single SWCNT. Our hydrogel platform is shown to significantly reduce the spectral heterogeneity of long SWCNTs in the hydrogel platform. Finally, we show the heterogenous spectral response on a single long SWCNT after the addition of calcium chloride.

632681 Diffusive Heat Transfer in Isolated, Free-Standing, Single-Walled Carbon Nanotubes

Abstract:

Carbon nanotubes show potential for thermal devices and heat management applications, albeit their thermal properties often remain poorly characterized. For example, room temperature axial thermal conductivity values reported from experiments span at least 5 orders of magnitude. This significant spread is partially due to challenges in accurately quantifying the different contributions to heat transfer at the nanoscale as well as in correctly identifying the type of CNT under study. Using Raman spectroscopy, we here measure isolated, single-walled CNTs in vacuum, reducing heat transfer pathways to the minimum. We show that a simple heat transfer equation based on Fourier's law can be applied to determine the axial thermal conductivity as only free parameter. We quantify the spread in this parameter extracted from repeated measurements on multiple free-standing segments of the same chirality tube to be on the order of 30 %. We attribute this variation to differences in the concentration of defects that act as scattering sites for phonons. A further reduction in thermal conductivity is observed upon filling the tube interior with water. Taking these extrinsic effects into account using Matthiessen's rule, we extract lower bounds for the intrinsic axial thermal conductivity of our single-walled CNTs typically on the order of several 100 W/(m K). The contactless platform discussed here represents a means to the statistically robust assessment of diffusive heat transfer in CNTs, laying the basis for more reliable systematic investigations on the thermal properties of these and related nanomaterials.

631409 Controlled Release from Intercalated Graphene Oxide Films: Edge and Basal-Plane-Specific Kinetics of Planar, 1D Wrinkled, and 2D Crumpled Nanochannels

Abstract:

Substances encapsulation and release are important techniques in many applications, including flavor, fertilizer, pesticide, and drug deliveries. The matching of release and demand is critical, which requires the release to be slow or controlled. In a controlled release system, substances are sustained in the specific matrix and released into desired media under physical or chemical control. Several methods have been demonstrated to obtain controlled release vehicles using various 2D materials. This research shows the edge- and basal-plane-specific kinetics of planar, 1D wrinkled, and 2D crumpled nanochannels of graphene oxide films used to control the release rates of molecular intercalants pre-loaded into GO gallery spaces. We show the study on rhodamine B dye, used as a model, show diffusive release rates in topography-related order. A pure RhB coated polymer substrate shows that approximately 80% of RhB was released in the first minute. RhB/GO intercalated. RhB within GO was slowly diffused; UV absorption shows a gradual increment of RhB in solution through 48 hrs, which reached equilibrium in 24 hrs. Here we will present the release rates and underlying pathways, including perpendicular and parallel to wrinkles. Moreover, we show the transport phenomena release profile of GO and citric acid intercalated, which gives diffusivity coefficients of $9.92*10^{-10}$ mm^2/min for planar films, $1.07*10^{-10}$ mm^2/min for 1D wrinkles, and 2D $5.53*10^{-10}$ mm^2/min for 2D crumple. This type of fluidic-space manipulation should allow the intelligent design of 2D-material-based technologies such as time-release drug-eluting coatings.

632866 Nanoporous Atomically Thin Graphene Membranes

Abstract:

Atomically thin 2D materials offer transformative opportunities for molecular separations. Pristine monolayer
graphene and h-BN are impermeable to species larger than protons but the introduction of nanoscale defects
in the 2D material lattice allows for the creation of size-selective nanoporous atomically thin membranes.
Here, I will discuss advances in 2D material synthesis and integration/processing routes to realize i) large area
atomically thin gas barriers, ii) fully functional nanoporous atomically thin membranes for dialysis based molecular separations, iii) novel approaches for in-situ growth of nanopores in 2D materials, iv) the
development of methods to probe sub-nanometer to nanometer defects over centimeter scale single crystalline 2D materials, v) roll-to-roll approaches to fabricate large area atomically thin membranes and vi)
atomically thin desalination membranes. Specifically, I will focus on the role of defects and associated
engineering challenges with quality and scalability for electronics vs membrane applications. Finally, I will
discuss our efforts to move these inventions from the lab space to the commercial arena.

References

Cheng P. et al. Nanoscale 2021

Cheng P. et al. NanoLetters 2020

Kidambi P.R. et al. Adv. Mat. 2018

Prozorovska L. and Kidambi P.R. Adv. Mat. 2018

Kidambi P.R. et al. ACS App. Mat. & Int. 2017

Kidambi P.R. et al. Adv. Mat. 2017

Kidambi P.R. et al. Adv. Mat. 2017

Kidambi P.R. et al. Nanoscale 2017

Kidambi P.R. et al. Chem. Mat. 2014

Kidambi P.R. et al. Nano Letters 2013

622979 Gold Nanoparticle-Enhanced siRNA Silencing in Plants Can Occur Independent of NP Internalization

Abstract:

Efficient delivery across a plant's cell wall and cell membrane remains a key challenge to the transport of biomolecular cargo into plant cells. Nanoparticle (NP)-mediated delivery of DNA and RNA to plants is an increasingly studied alternative to current delivery strategies[1–3], however, little is known about how the size and shape of nanoparticles influences their transport in plants and the resulting efficiency of cargo delivery.

This rapidly growing interest in nanoparticle-mediated delivery of DNA and RNA to plants requires a better understanding of how nanoparticles and their cargoes translocate in plant tissues and into plant cells. Here, we employ non-biolistically delivered DNA-modified gold nanoparticles (AuNP) spanning various sizes (5 – 20 nm) and shapes (spheres and rods) to systematically investigate their transport following infiltration into Nicotiana benthamiana (Nb) leaves using both confocal microscopy and transmission electron microscopy. Generally, smaller AuNPs demonstrate more rapid and higher levels of association with plant cell walls compared to larger AuNPs. We observe internalization of rod-shaped but not spherical AuNPs into plant cells, yet surprisingly, 10 nm spherical AuNPs functionalized with small-interfering RNA (siRNA) are most efficient at siRNA delivery and inducing GFP gene silencing in mature transgenic Nb plant leaves.

Based on our results, we propose a 'reservoir' mechanism in which NPs embed in the cell wall, protecting the siRNA from degradation, whilst slowly releasing cargo into plant cells to achieve gene silencing. Contrary to initial expectations, our results demonstrate that NP-mediated cargo delivery can be achieved without NP internalization into plant cells. These results indicate the importance of nanoparticle size in efficient biomolecule delivery, and, counterintuitively, demonstrate that efficient cargo delivery is possible and potentially optimal in the absence of nanoparticle cellular internalization. Our results highlight nanoparticle features of importance for transport within plant tissues, providing a mechanistic overview of how nanoparticles can be designed to achieve efficacious bio-cargo delivery for future developments in plant nanobiotechnology.

References:

(1) Demirer, G. S.; Zhang, H.; Matos, J. L.; Goh, N. S.; Cunningham, F. J.; Sung, Y.; Chang, R.; Aditham, A. J.; Chio, L.; Cho, M. J.; Staskawicz, B.; Landry, M. P. High Aspect Ratio Nanomaterials Enable Delivery of Functional Genetic Material without DNA Integration in Mature Plants. *Nat. Nanotechnol.* **2019**, *14* (5), 456–464. https://doi.org/10.1038/s41565-019-0382-5.

(2) Mitter, N.; Worrall, E. A.; Robinson, K. E.; Li, P.; Jain, R. G.; Taochy, C.; Fletcher, S. J.; Carroll, B. J.; Lu, G. Q.; Xu, Z. P. Clay Nanosheets for Topical Delivery of RNAi for Sustained Protection against Plant Viruses. *Nat. Plants* **2017**, *3* (2), 1–10. https://doi.org/10.1038/nplants.2016.207.

(3) Torney, F.; Trewyn, B. G.; Lin, V. S. Y.; Wang, K. Mesoporous Silica Nanoparticles Deliver DNA and Chemicals into Plants. *Nat. Nanotechnol.* **2007**, *2* (5), 295–300. https://doi.org/10.1038/nnano.2007.108.

629739 Developing Thin, Drug-Eluting Topical Ocular Gels Formed By Administration of Low Concentration, Thermoreversible Polymer Solutions in Hypotonic Aqueous Vehicles

Abstract:

Problem: Eye drops are the most common ophthalmic pharmaceutical dosage form for treating and preventing ocular diseases. However, it is estimated that <5% of drugs dosed topically will be absorbed into the eye, while the rest is rapidly removed from the ocular surface. Therefore, many eye drops must be dosed multiple times per day to reach therapeutics levels, which unfortunately reduces patient compliance. Here, we engineered an *in situ* gelling hypotonic eye drop formulation that form a uniform, thin, and clear gel layer that resides longer on the ocular surface and delivers the drugs to the eye for a prolonged period of time.

Methods: Pluronic F127 (aka Poloxamer 407) was selected due to thermosensitive gelling properties and use in FDA-approved products. Multiple-particle tracking (MPT) was used to assess *in vivo* gelation properties. Fluorescently-labeled 200 nm PEGylated nanoparticles were added to eye drop formulations (0.0004% w/v) and dosed (5 µL) topically to Sprague Dawley rat eyes ($n = 4\text{-}7$ per group). An epifluorescence microscope were used to take time-lapse videos to calculate the ensemble-averaged mean square displacement (MSD). For the intraocular pressure (IOP) lowering efficacy study, the hypotonic gel-forming eye drop was formulated containing 0.15% w/v brimonidine tartrate (12% Hypo BT) and compared to both conventional thermogels formulated at the CGC (18% Iso BT) and commercially available Alphagan® P 0.15% eye drops. A single dose (50 µL) was topically administered to the eye of normotensive rabbits ($n = 5\text{-}6$ per group) and the IOP were measured with hand-held rebound tonometer (TonoVet) at various times up to 10 h. For posterior pharmacokinetics studies, the hypotonic gel-forming eye drop was formulated containing 0.4% w/v sunitinib malate (12% Hypo SM). Juvenile Yorkshire pigs (n = 4) were dosed daily with 12% Hypo SM for 5 days. The retina and choroid were collected 1 h after the last dose, and drug concentrations were measured by liquid chromatography–tandem mass spectrometry.

Results: We discovered that by formulating the polymer below the critical gel concentration (CGC) with ion content lower than physiological conditions, a thin, uniform gel layer could be formed over the ocular surface. The *in vivo* MPT study confirmed that the hypotonic 12% F127 solutions formed a gel structure on the ocular surface as evidenced by the low particle MSD, similar to the conventional 18% F127 gel. In the IOP lowering efficacy study, we demonstrated that 12% Hypo BT formulation provided the most significant and sustained decrease in IOP, which was still significant compared to the Alphagan® P ($p < 0.0036$) and 18% isotonic BT ($p < 0.0012$) at 10 h. Finally, we sought to test if the hypotonic gel-forming eye drop could deliver therapeutically relevant drug concentrations to the posterior segment in larger animals with eyes more similar in size to human eyes. The drug concentrations in the retina (140 ng/g) and choroid/retinal pigment epithelium (226 ng/g) were well above the Ki for SM inhibition of vascular endothelial growth factor receptor and platelet-derived growth factor receptor after topical dosing with 12% Hypo SM for 5 days.

Implications: Here we describe an innovative discovery for formulating thermosensitive *in situ* gelling polymers for enhanced topical ocular drug delivery. We characterized the 12% hypotonic gelling formulation properties and demonstrated more sustained IOP lowering compared with commercial Alphagan® P formulation. We further observed that the 12% hypotonic gelling formulation provided therapeutically relevant drug delivery to the posterior segment in large animals, demonstrating the potential for posterior segment delivery with an eye drop in a human-sized eye. Longer lasting and more sustained ocular drug delivery can reduce eye drop dosing frequency and improve patient quality of life.

633086 Award Submission: Development of Stable Targeted Nano-, Encapsulated Manganese Oxide (NEMO) Particles for Early Breast Cancer Diagnosis By MRI

Abstract:

The ultimate goal of this research project is to develop a novel contrast agent to reduce the number of false-positive and false-negative imaging results associated with current breast cancer imaging techniques. Gadolinium chelates are the current clinically approved contrast agents for breast magnetic resonance imaging (MRI), but they are always "on" and highlight any vascularized structure. Due to their lack of targeting, both benign and malignant breast tumors are enhanced with gadolinium, resulting in high false-positive rates up to 25%. By utilizing tumor-targeted pH-sensitive manganese oxide (MnO) nanoparticles, a contrast agent can be developed that will only turn "on" after internalization into cancer cells and dissolution within low pH endosomes/lysosomes.

Nanocrystalline MnO was synthesized through thermal decomposition of Mn(II) acetylacetonate in dibenzyl ether and oleylamine at 280°C. After confirmation of nanocrystalline MnO size and chemistry with TEM, XRD, and FTIR, MnO nanocrystals were encapsulated within poly(lactic-co-glycolic acid) (PLGA) and poly(ethylene glycol) (PEG) through single emulsion to form Nano-, Encapsulated, Manganese Oxide (NEMO) particles. Particles with different PEG percentages (2.5, 5, and 10% w/w) and PLGA terminations (carboxylic acid vs. ester) were synthesized and incubated in water over 24 hours to assess particle stability. The hydrodynamic diameter and charge of NEMO particles were measured pre- and post-incubation with dynamic light scattering (DLS) and zeta potential, respectively. For carboxylic acid terminated PLGA, a higher PEG % resulted in enhanced particle stability due to a more negative zeta potential that favored particle-particle repulsion (2.5% PEG: -12.89 mV; 5% PEG: -14.31 mV; 10% PEG: -16.33 mV). Surprisingly, increased PEG % had the opposite effect with ester terminated PLGA; particles became less stable with higher PEG due to a less negative zeta potential that reduced interparticle repulsion (2.5% PEG: -16.51 mV; 5% PEG: -12.61 mV; 10% PEG: -8.557 mV). As higher PEG density can reduce nanoparticle cell uptake, NEMO particles with 2.5% PEG ester terminated PLGA were chosen as the optimal particle for further experiments.

To enable specific uptake by cancer cells, a tumor-targeting peptide against underglycosylated mucin-1 (uMUC-1) was attached through click chemistry using copper (I) as the catalyzer. The uMUC-1 targeting peptide conjugated NEMO particles were characterized with DLS and a fluorometric assay to assess hydrodynamic size (~180 nm) and correct targeting attachment, respectively. For NEMO particles to cause a positive T_1 signal on MRI, MnO must dissolve and release Mn^{2+} at low pH. To evaluate the Mn^{2+} release and MRI properties, NEMO particles were incubated at 3 different pHs over time including pH 7.4 (blood pH), 6.5 (pH of tumor extracellular space), and 5 (endosome/lysosome pH), and supernatants were evaluated using inductively coupled plasma-optical emission spectroscopy (ICP-OES) and 1T MRI. NEMO particles released the most Mn^{2+} at pH 5 (~93% at 24 hr), which produced the lowest T_1 value with the brightest MRI signal at the first two time points (848 ms at 1 hr, and 886 ms at 2hr, pH 5). The substantial MRI contrast generated after just 1-2 hours indicates that NEMO particles can result in a measurable contrast change within clinically relevant time frames.

To assess *in vivo* stability and toxicity, different dosages of tumor-targeted NEMO particles (15, 7.5, 3, and 1.5 mg Mn/kg) were injected once a week for three weeks into BALB/c female mice. Weight, behavior, and appearance were scored daily to measure toxic effects. Only the highest dosage, 15 mg Mn/kg, had an impact on mouse weight during the first 24 hr (~1 g weight loss), and resulted in a slight decrease in natural and provoked behavior; mice regained weight and normal behavior within 2 to 3 days post-injection. Based on these results, overall NEMO particles were well tolerated *in vivo* over 3 weeks. In future studies, 7.5 mg Mn/kg will be utilized to maximize Mn dose for enhanced MRI signal while avoiding the weight and behavicral effects associated with 15 mg Mn/kg. Future work includes further *in vitro* studies of NEMO particle cell labeling and toxicity, along with *in vivo* MRI of tumor-bearing mice to assess the specific detection of malignant versus benign breast tumors with NEMO particles.

632250 Graduate Student Award Session: Liposome and Polyelectrolyte Layers Derived Single Shot Vaccine Platform for Controlled Release of Inactivated Chikungunya Virus

Abstract:

Immunization through inactivated virus enhances the safety profile of the vaccine and is one of the most efficient technique to protect against infectious diseases. However, inactivated vaccines require multiple rounds of vaccinations that often lead to missing or mistimed doses thereby limiting their potency. This multidose requirement further limits mass immunization during the outbreak of an infectious disease such as chikungunya virus (CHIKV) in a place without herd immunity. Single shot vaccines capable of releasing the antigen intermittently can potentially be employed to overcome the shortcomings of multi-dose vaccines and provide herd immunity in shorter period of time. A sustained release of antigen can help in maintaining high concentration of antibodies in the system and programmed delayed release can induce strong booster effect. In this study, we engineered a single shot vaccine delivery platform comprising liposomes (LNP) embedded polyelectrolyte multilayer (PEM) lattice to provide controlled spatiotemporal release of CHIKV antigen and thus induce long term immunity. The CHIKV virus inactivated using γ-radiations in presence of MDP complex (MDP-iCHIKV) was injected in mice and tested for production of antibodies. The MDP-iCHIKV were then encapsulated in LNP followed by embedment in PEM films so as to obtain a programmed release of antigen at different time points. We performed the characterization of LNP and LNP-PEM assembly using dynamic light scattering method to determine the size and zeta potential measurements to determine the surface charge. Further confocal imaging and scanning electron microscopy was conducted to analyze the structure properties of the LNP-PEM assembly. Analysis of encapsulation and released CHIKV antigen concentration was performed using ELISA. In-vivo studies were performed by injecting the LNP-PEM assembly in mice followed by quantification of anti-CHIKV IgG in serum at different time points. The studies showed that MDP complex protected CHIKV viral envelop proteins while destruction of its genome by γ-radiation and also induced anti-CHIKV IgG response upon injection in mice. Significant changes in size, charge and structural properties were observed in the process of LNP embedment into PEM thus confirming the intended design of the vaccine plaform . We obtained about 75% encapsulation efficiency of CHIKV antigen in LNP and nearly 80% adhesion efficiency of LNP on PEM. In-vivo studies showed patterned increase and decrease in production of anti-CHIKV IgG indicating towards intermittent release of antigen from the LNP-PEM assembly. We are further working on optimizing the design of this single-shot vaccine platform to achieve a programmed release of antigen at multiple stages. We believe that this design for single shot vaccine platform would not only eliminate the need for multiple booster shots for inactivated vaccines but can also be potentially employed for long term immunization against a variety of pathogens.

627931 Near-Infrared Catecholamine Nanosensors Reveal Disruptions in Dopamine Release in Huntington's Disease Mouse Models

Abstract:

Dopamine neuromodulation of neurotransmitter signaling is a critical process that facilitates learning, motivation, and motor control. Disruption of these processes has been implicated in a number of neurological and psychiatric disorders including Huntington's Disease (HD), a neurodegenerative disorder characterized by the production of mutated huntingtin protein and a triad of motor, behavioral, and cognitive symptoms. Despite the established clinical relationship between HD and dopamine, the exact mechanisms through which disease develops and progresses is not fully understood. Similarly, though present treatments for HD symptom management often broadly target dopamine neuromodulation, their exact mechanism of action remains elusive owing to challenges in studying dopaminergic neuromodulation. In contrast to classical neurotransmitter signaling, dopamine neuromodulation exerts it influence by diffusing beyond the synaptic cleft to influence the excitability of neighboring neurons. As such, studying the release and spread of dopamine at the spatio-temporal resolution of single dopamine release sites (μm, ms) and beyond is critical to not only understanding HD disease mechanisms but also understanding how to best develop novel therapies (RNAi, cell replacement) to target HD affected neural circuits.

To this end, we employ near-infrared fluorescent catecholamine nanosensors (nIRCats) to image dopamine release within the brain striatum of R6/2 Huntington's Disease Model (R6/2) mice. We have previously shown that these near-infrared fluorescent, polymer-functionalized semiconducting single wall carbon nanotubes serve as adept dopamine sensors in the basal ganglia, exhibiting $\Delta F/F$ of up to 24-fold in the fluorescence emission window of 1000-1300 nm, a wavelength ideal for imaging in optically scattering brain tissue[1]. In this study, we find that stimulated dorsal striatal dopamine release decreases with progressive degeneration of motor ability, consistent with previous trends identified through microdialysis and Fast Scan Cyclic Voltammetry. Notably, nIRCats's high spatial resolution allows further elucidation of this process, revealing that these decreases are primarily driven by a decrease in the number of dopamine releasing sites rather than a decrease in individual release site performance. We further interrogate this late disease state, by examining the external calcium sensitivity of these dopamine release sites and utilizing nIRCats' compatibility with dopamine pharmacology to assess changes in Dopamine D2 Receptors (D2R) expression. These findings, enabled by nIRCats, provide a more detailed look into how dopamine release is disrupted and dysregulated during Huntington's Disease. Furthermore, they provide a framework to understand the brain during disease and underscore the utility of nIRCats as a versatile new optical tool for dopamine detection in the brain.

1. Beyene, A. G., Delevich, K., **Yang, S.J.** et al. Imaging Striatal Dopamine Release Using a Non-Genetically Encoded Near-Infrared Fluorescent Catecholamine Nanosensor. *Science Advances* (2019).

630494 Award Submission DNA Origami Tubes with Reconfigurable Cross-Sections

Abstract:

DNA nanotechnology has emerged over the last decades as a promising approach for various biological, mechanical, and environmental applications. The programmable nature of DNA bases provides a framework to construct nanoscale devices with complex geometry, precise motion, and controlled dynamic and mechanical properties. Dynamic devices in particular have emerged as promising tools for various biological, mechanical, and environmental applications such as biophysical measurement devices, nanorobotic components, triggered drug delivery vehicles, and nanobiosensors. The unique properties of DNA nanostructures also make them highly attractive for bottom-up assembly of materials with dynamically programmable properties. Hence, coupling across length scales, i.e. using individual devices on the molecular scale to make larger assemblies including microscale robotic systems and materials, is an exciting avenue for DNA nanotechnology. There have been major advances over the last decade in higher order assembly of DNA structures, including recent efforts to make reconfigurable arrays and lattices; however, the dynamic behavior of these assemblies remains limited compared to the major advances in complexity of individual devices. Here our goal is to provide an approach using dynamic DNA devices capable of reconfigurable higher order assembly to advance the complexity of reconfigurable arrays and lattices on a larger scale.

We leverage scaffolded DNA origami (DO), which is a molecular self-assembly approach where many short single-stranded DNA (ssDNA) oligos, referred to as "staples," bind with a long ssDNA strand of a known sequence called a "scaffold." The staples bind to the scaffold using Watson-Crick complementary base-pairing in precise locations to create double-stranded DNA (dsDNA) regions that fold into precisely designed shapes. Early work on DO structures focused on static structures and shapes, whereas more recent work on dynamic DO structures include motion or actuation to increase complexity. Advanced 3D dynamic devices include objects with 1D motion such as hinges or pistons, and mechanisms with more complex motion such as crank-sliders or Bennet linkages with programmed kinematic motion or actuation capabilities. A number of recent efforts have expanded DO to larger scales through hierarchical assemblies that integrate many individual structures in a controlled way, and prior efforts have demonstrated integration of reconfigurable constructs into arrays. However, the reconfiguration of higher order assemblies remains quite simple to 1D rotation and only open-to-close reconfiguration. Therefore, we developed a reconfigurable 6-component DNA origami mechanism, the 6-bar, that can be reconfigured into multiple shapes (Figure 1A) and assembled into linear arrays (Figure 1C) with different cross-sections. Our goal is to demonstrate the reconfiguration of DO shapes into a variety of stiff cross-sections and demonstrating the reconfiguration of DO tubes between several distinct cross-section shapes.

The 6-bar consists of 6 arms or links, consisting of 12 double-stranded DNA (dsDNA) helices arranged in a 3x4 square lattice cross section. These 6 bundles are connected via ssDNA hinges to form a closed-loop mechanism. The "free" state is flexible and contains 172 nucleotide (nt) scaffold loops on each bundle. These scaffold loops are present to enable the formation of struts that connect neighboring components at a specific angle to determine the shape of the mechanism. The struts can create bundle angles of 60°, 90°, 120°, and 180° (Figure 1A). Each folded configuration contains three full struts, and all the staples comprising the struts contain overhanging ssDNA extensions to serve as toe-holds for toe-hold mediated strand displacement (TMSD). Each configuration can thus reconfigure back to the free (i.e. no strut) state with the use of TMSD, and then to a different geometry through the addition of new strut strands. Figure 1B illustrates a transition between an open rectangle to a flat closed geometry with high efficiency of reconfiguration. Reconfiguring the individual 6-bar devices provides a basis for reconfiguring higher-order assemblies. We expand the reconfiguration to polymer "tubes" with the same reconfigurable cross-sections. We can assemble these DO polymers into several different cross-section shapes (Figure 1C). These polymers allow us to bridge the gap between individual device reconfiguration to multi-device reconfiguration.

In addition to demonstrating reconfiguration of individual devices and tubes, the 6-bar provides an interesting platform to study free energy differences of the various conformations that only vary by the composition of the struts. Introducing different sets of strut staples to the free (i.e no strut) state allows us to analyze the preferred, or stable, configurations. Exhibiting reconfiguration between multi-states also allows us to study the energetics between the open to closed loop structures. We expanded this idea to structural tuning of the interactions as an approach to control formation of higher order assemblies such as tubes with varying cross-sections across the length. Studying the energetics of the higher order assemblies may provide new ways to control assembly to create more complex reconfigurable DO structures.

This work expands upon molecular motors and machines to include nanoscale and micron-scale mechanisms. The 6-bar mechanism and 6-bar polymers could be used to organize components over a micron length-scale or create assemblies with inhomogeneous properties. DNA origami includes many applications such as drug delivery, signal transmission, and biological mechanisms, however has expanded broadly to many different fields outside of biology. Using DNA as a material to construct mechanisms studied in fields such as chemical, mechanical, and bio-engineering has been an amazing advancement provided the biological nature and application of DNA.

Figure 1: **A** (left) Course-grain molecular dynamic simulations of 6-bar mechanism. (right) Transmission electron microscope (TEM) image averages of each 6-bar configuration (scale bar = 50nm). **B** (left) Reconfiguration schematic. (right) TEM images at each step of reconfiguration with image average insets (scale bar = 50nm). **C** (top) 6-bar polymer schematics and (below) TEM images of 6-bar polymers (scale bar = 100nm).

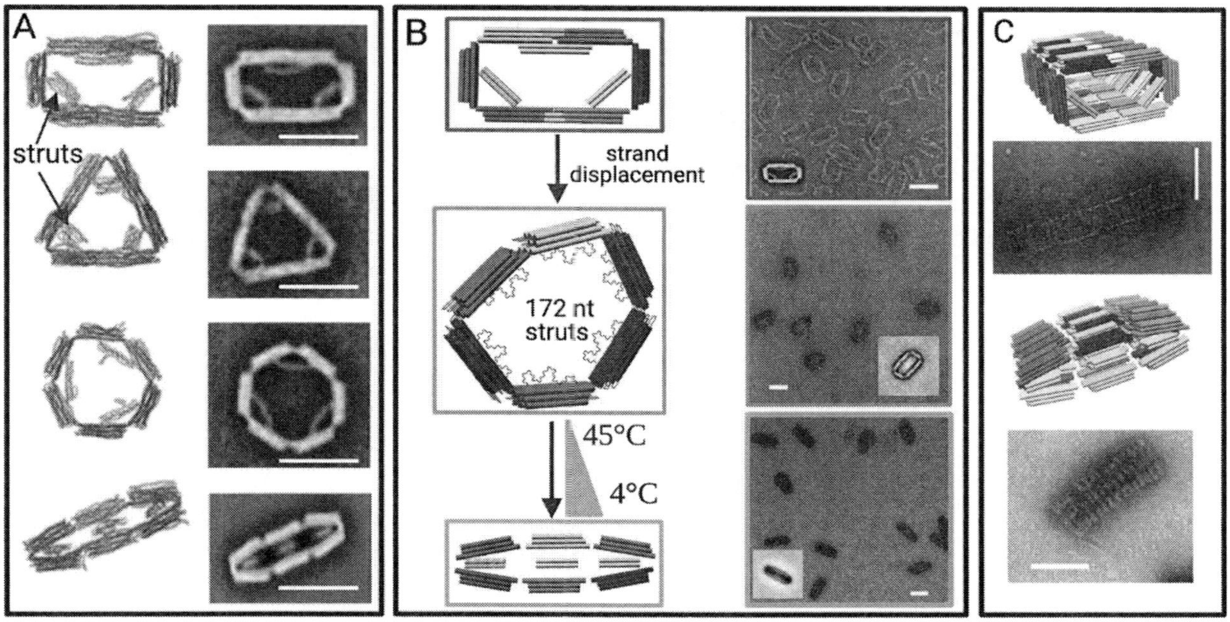

630948 Nanoconfined Water Endows Peptidoglycan Extreme Water-Responsive Actuation

Abstract:

Water-responsive (WR) materials that deform in response to changes in relative humidity (RH) show great potential in applications, including soft robotics and novel energy harvesting systems. The best-performance WR material up-to-date is *Bacillus (B.) subtilis* spore, which shows an actuation energy density of ~10 MJ m^{-3}, surpassing that of existing artificial muscles and actuators. However, the underlying reason of spores' forceful actuation remains unclear. Here, we present that peptidoglycan (PG) within spores' cortex layer dominates spores' WR behaviors, and its WR energy density reaches 59.9 MJ m^{-3}, which exceeds spores' record-high energy density by three-fold. We found that, when RH is increased from 10% to 90%, spore PG exhibits a WR strain of ~50.1%, which is ~5 times larger than that of spores. We also found that the PG from *B. subtilis* bacterial cell shows comparable water-responsiveness. For example, cell wall PG shows a WR strain of 27.2% and an energy density of 72.9 MJ m^{-3}. To elucidate water's effect on PGs, we analyzed the effect of water's diffusion on the PG solid network and found that the viscosity of water in PG is extremely high (~7 orders higher than bulk water), which enables the water flow to drag and deform PG's solid network. Our finding of PG's extreme water-responsiveness not only provides a new record energy density for artificial muscles, but also provides mechanistic insight in developing and exploring more advanced WR materials.

630898 Polypeptide Templating of Structural Proteins for Tailored Hierarchical Materials

Abstract:

Materials generation in nature often involves the use of templates to regulate phase transformation and subsequent hierarchical organization of nanoscale building blocks, with the most prominent examples being biomineralization and amyloid fibrils growth, to name a few. Despite recent progress in directing the assembly of biomacromolecules into well-defined nanostructures, leveraging pathway complexity of molecular disorder to order transition while bridging materials fabrication from nano- to macroscale remains an open challenge. In this presentation, we introduce templated crystallization of structural proteins as a process to nanofabricate hierarchically structured materials up to centimeter scale, using silk fibroin as an example. The process involves the use of ordered peptide supramolecular assemblies as templates to drive a phase transformation of silk fibroin from unordered to ordered conformations, thereby enabling further assembly of the silk fibroin chains into nanostructured materials (i.e. β-sheeted nanofibrils). Multiple parameters including the relative concentration between silk fibroin and peptide seeds, silk fibroin molecular weight, pH and peptide conformation are investigated to fine tune the assembly kinetics and silk polymorphs. Templated crystallization coupled with various top-down fabrication such as soft lithography and printing enables (i) epitaxial materials growth from the molecular level (i.e. disordered silk fibroin molecules) all the way up to centimeter scale in an exquisite single-pot, bottom-up system; (ii) surface functionalization with topographical control over nanofibrils assembly on pre-deposited peptide seeds; and (iii) three-dimensional manipulation of silk nanofibrils into macroscopic structures with customized shapes and controlled anisotropy. Together, these results pave the way for nanofabrication of a new generation of smart and adaptive materials built from the bottom up.

620665 Synthesis and Characterization of Biogenic Selenium Nanoparticles Made from Pathogenic Bacteria with Selective Antimicrobial Properties

Abstract:

Antimicrobial Resistance (AMR) to antibiotics has become an urgent crisis for humanity, incurring both medical and economic burden into a continuously growing society. Making matters worse, the future does not offer a relief, as the current overuse and misuse of antibiotics contribute to the rapid development of novel resistant strains of pathogens that we will need to eventually fight. Thus, an alternative, non-resistant-inducing and effective way to treat infection is needed. A perfect candidate comes from the nanoscale and the use of metal-based nanoparticles (MNPs), systems which have long been studied for their antimicrobial properties. However, the cost, efficiency, processing difficulties and ecological safety of traditionally synthesized MNPs prevent their clinical translation. To remedy these challenges, biogenic synthesis of nanoparticles, using living systems (bacteria or fungi, among others), allow for a potentially successful implementation. Among all the possible candidates, selenium nanoparticles (SeNPs) are explored and presented as an alternative antimicrobial agent without the associated resistance problems found in silver nanomaterials.

In this work, SeNPs were synthesized using *Escherichia coli (EC)*, *Pseudomonas aeruginosa (PA)*, *Staphylococcus aureus (SA)* and Methicillin-resistant *Staphylococcus* aureus (*MRSA*), with a diameter of 80-120 nm by culturing bacteria cells in standard conditions and inoculating them with a selenium salt at the end of exponential growth phase. The nanostructures were characterized using Transmission Electron Microscopy (TEM) and Energy Dispersive X-Ray (EDX) to determine the chemical compositions and Fourier transform infrared spectroscopy (FTIR) to validate the chemistry within the samples. Each type of the synthesized SeNPs were tested in antimicrobial studies against the same species that produced them (homogeneous treatment) and other species (heterogeneous treatment) for their ability to inhibit the bacterial growth through optical density measurements and colony forming unit assays. Biocompatibility tests of the SeNPs with human dermal fibroblasts (HDF) were completed, along with resistance studies performed as several rounds of exposure onto pathogens.

Results indicate that SeNPs were successfully generated by all tested strains and exhibited an organic coating layer identified as the protein corona, with unique composition depending on host biofactory. The biogenic SeNPs exhibited antimicrobial properties, with log reduction ranging from 3 – 4 in heterogeneous applications and 5 – 7 in homogeneous applications. Cytotoxicity assays confirmed its safety profile against HDF cells. More significantly, biogenic SeNPs treatment did not induce resistance, in contrast with penicillin and commercialized AgNPs.

These findings confirm the viability of bacteria as a synthesis pathway of SeNPs, and in extension, other metallic NPs. Generated biogenic SeNPs exhibit desired size and antimicrobial properties, with enhanced effectiveness when applied homogeneously. This specificity, along with its limited resistance-inducing property, open the door for biogenic SeNPs to become an easy, safe, cost-effective and potent treatment and a potential solution against the AMR crisis.

Figure: SEM images of MDR-E.coli (A,B) and MRSA (C,D) producing SeNPs, along with TEM images of SeNPs produced by E.coli (E) and S.A (F). Antimicrobial activities of SeNPs produced by E.coli against several pathogens (G) and resistance study (H) demonstrates selectivity and non-resistant-inducing properties

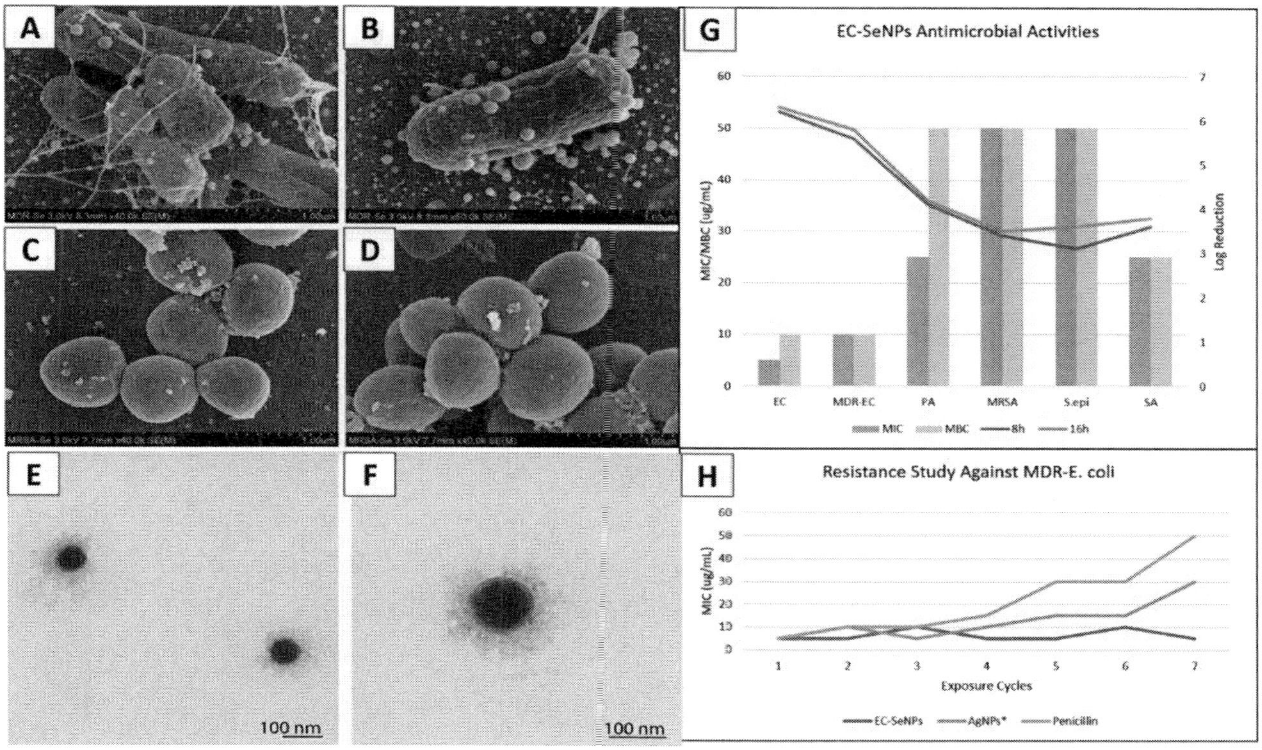
A
B
C
D
E
F
G
H
100 nm
100 nm
EC-SeNPs Antimicrobial Activities
MIC/MBC (ug/mL)
Log Reduction
EC
MDR-EC
PA
MRSA
S.epi
SA
MIC
MBC
8h
16h
Resistance Study Against MDR-E. coli
MIC (ug/mL)
Exposure Cycles
EC-SeNPs
AgNPs*
Penicillin

AIChE Abstract #622677

Single Particle ICP-MS: An Emerging Technique for Quantifying Size and Aggregation of Inorganic Nanomaterials for Biomedical Applications

Nathan Donahue, Emmy Francek, Emi Kiyotake, Emily Thomas, Wen Yang, Lin Wang, Michael Detamore and Stefan Wilhelm, Stephenson School of Biomedical Engineering, The University of Oklahoma, Norman, OK

Abstract Text:
The efficacy, toxicity, and fate of engineered nanomaterials depends on size and colloidal stability. Conventional techniques to characterize the size and aggregation of nanomaterials are limited to batch-based measurements or require harsh conditions to obtain single nanoparticle resolution. Single particle inductively coupled plasma mass spectrometry (SP-ICP-MS) is an emerging technique that offers high-throughput *in situ* nanoparticle size analysis with single nanoparticle resolution.

To accomplish this, SP-ICP-MS utilizes a microfluidics sample introduction system to introduce nanoparticles—one at a time—into an argon plasma. In the argon plasma, individual nanoparticles are atomized and ionized to generate an ion cloud. This discrete ion cloud then travels through a quadrupole mass spectrometer. Once detected, the measured intensity of a singular ion cloud is used to calculate the size of individual nanoparticles. This process occurs at a rate of +200 nanoparticles per minute for a given element.

We show that SP-ICP-MS accurately determined size distributions of different sizes of gold nanoparticles (AuNPs) that we synthesized in house at a rate of ≥ 200 particles/minute. Dynamic light scattering and transmission electron microscopy corroborated our SP-ICP-MS results. After successfully determining the mass and size distributions of AuNPs with different sizes, we used SP-ICP-MS to differentiate between nanoparticles of different sizes in mixtures. Our results confirmed that SP-ICP-MS can simultaneously and precisely measure AuNP mass distributions of multiple nanoparticle subpopulations in a nanoparticle mixture with single particle resolution. To further validate SP-ICP-MS, we demonstrated that SP-ICP-MS can accurately determine the mass distribution of non-spherical anisotropic nanoparticles.

Since SP-ICP-MS differentiated subpopulations of nanoparticles in mixtures and quantified the mass distribution of anisotropic nanoparticles, we hypothesized that SP-ICP-MS could quantify nanoparticle aggregation with single particle (i.e., single aggregate) resolution. To generate AuNP aggregates, we exposed citrate-coated 16-nm AuNPs with a narrow size distribution to physiologically relevant saline concentrations. Given that the mean mass of our 16-nm AuNPs was 50 attograms, we assumed that 100 attograms corresponded to an aggregate consisting of 2 AuNPs, 150 attograms corresponded to an aggregate consisting of 3 nanoparticles, and so forth.

SP-ICP-MS detected nanoparticle aggregation within 5 minutes of saline exposure. After 60 minutes of saline exposure, the number of individual single AuNPs decreased by ~70% when compared with AuNPs in the initial measurement without saline. At this time point, more than half (52%) of the aggregates were comprised of 2–5 AuNPs per aggregate while only ~ 21% of the detected aggregates consisted of 5 or more AuNPs per aggregate.

Our SP-ICP-MS experiments confirmed the aggregation data obtained from dynamic light scattering and UV-Vis spectrophotometry which showed rapid progression of salt-induced AuNP aggregation. However, in stark contrast to batch-based methods, we quantified the individual masses of AuNP aggregates. When compared with light scattering techniques, SP-ICP-MS results were not skewed by larger nanoparticles or by the extra layer of hydration, as we only measure gold ions from the AuNPs. When compared to electron microscopy, SP-ICP-MS circumvents harsh sample preparation conditions that may introduce artifacts and the need for subsequent image analysis. Ultimately, SP-ICP-MS is an unbiased approach for quantifying nanoparticle aggregation *in situ*.

To mitigate nanoparticle aggregation, we hypothesized that PEGylation of AuNPs, —the decoration of nanoparticle surfaces with polyethylene glycol (PEG) polymers—could be adopted for our study. We prepared 16-nm AuNPs modified with various surface densities of thiol-PEG5kDa-methoxy. Upon exposure to saline, SP-ICP-MS results confirmed that PEGylation of nanoparticles physically hinders the rapid onset of salt-induced aggregation in a PEG surface density–dependent fashion. After 60 min in saline, more than 90% of all detected nanoparticle events were detected as individual nanoparticles for each PEG surface density compared with the 27% of individual nanoparticles observed for citrate coated AuNPs. SP-ICP-MS results show that the formation of larger aggregates (3 or more nanoparticles per aggregate) was completely inhibited by PEG. Our SP-ICP-MS findings confirmed as little as 0.010 PEG/nm^2 was sufficient for reducing the time-dependent effects of aggregation for monodisperse 16-nm AuNPs.

In summary, we show that SP-ICP-MS accurately analyzed the size of engineered AuNPs with varying shapes and sizes. Furthermore, we applied SP-ICP-MS as an *in situ* technique to quantify aggregation of AuNPs in physiologically relevant saline conditions and showed the mass distributions of AuNP aggregates with single aggregate resolution. We applied surface engineering strategies to demonstrate that increasing PEG surface densities on AuNPs could block aggregation.

SP-ICP-MS analysis of inorganic nanomaterials has far reaching implications for researchers who seek to understand how engineered nanoparticles change over time in biologically relevant environments and how surface modifications impact nanoparticle transformations.

SP-ICP-MS to assess citrate-coated AuNP aggregation. **a** Mass histograms of citrate-coated 16-nm colloidal AuNPs exposed to 150 mM aqueous NaCl solution over time. **b** Detected mass distribution in panel a is converted into number of nanoparticles per aggregate (bin size = 50 ag). **c** Percentage of individual nanoparticles are compared with masses > 100 ag, a mass which is equal to > 2 nanoparticles per aggregate (i.e., detected event). **d** Breakdown of AuNP aggregates into bins of 2–5 and > 5 AuNPs per aggregate (i.e., detected event)

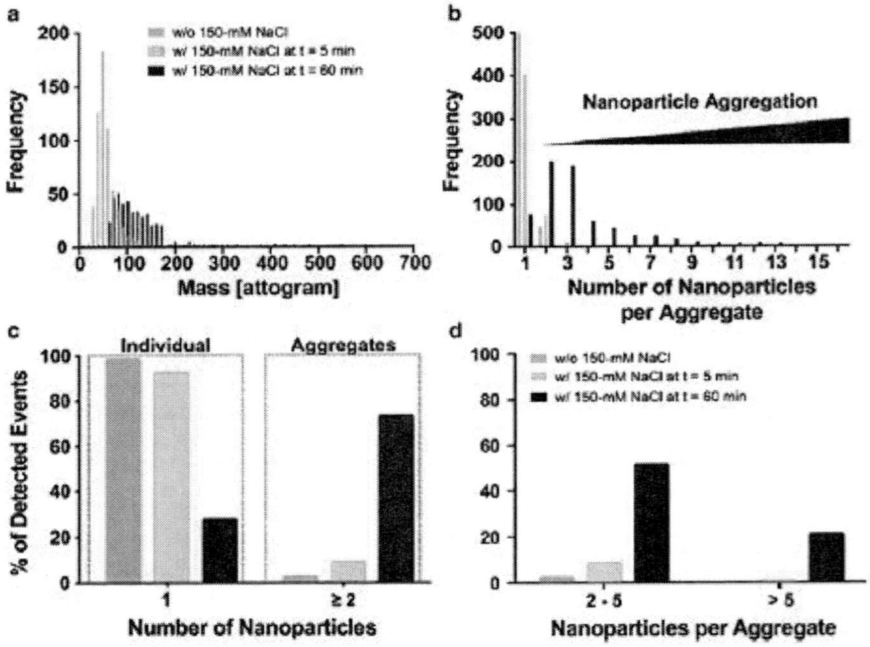

625120 Multispectral Fingerprinting Resolves Dynamics of Nanomaterial Trafficking in Primary Endothelial Cells (Award Session)

Abstract:

Intracellular vesicle trafficking involves a complex series of biological pathways used to sort, recycle, and degrade extracellular components, including engineered nanomaterials (ENMs) which gain cellular entry *via* active endocytic processes. A recent emphasis on routes of ENM uptake has established key physicochemical properties which direct certain mechanisms, yet relatively few studies have identified their effect on intracellular trafficking processes past entry and initial subcellular localization. Here, we developed and applied an approach where single-walled carbon nanotubes (SWCNTs) play a dual role - that of an ENM undergoing intracellular processing, in addition to functioning as the signal transduction element reporting these events in individual cells with single organelle resolution. We used the unique optical properties exhibited by non-covalent hybrids of single-stranded DNA and SWCNTs (DNA-SWCNTs) to report the progression of intracellular processing events *via* two orthogonal hyperspectral imaging approaches of near-infrared (NIR) fluorescence and resonance Raman scattering. A positive correlation between fluorescence and G-band intensities was uncovered within single cells, while exciton energy transfer and eventual aggregation of DNA-SWCNTs were observed to scale with increasing time after internalization. These were confirmed to be consequences of intracellular processes using pharmacological inhibitors of endosomal maturation, which suppressed spectral changes through two distinct mechanisms. An analysis pipeline was developed to colocalize and deconvolute the fluorescence and Raman spectra of subcellular regions of interest (ROIs), allowing for single-chirality component spectra to be obtained with sub-micron spatial resolution. This approach uncovered a complex relationship between DNA-SWCNT concentration, fluorescence intensity, environmental transformations, and irreversible aggregation resulting from the temporal evolution of trafficking events. Finally, a spectral clustering analysis was applied to delineate the dynamic sequence of processes into four distinct populations, allowing stages of the intracellular trafficking process to be identified by the multispectral fingerprint of encapsulated DNA-SWCNTs.

632015 Protein Corona Formation Studied with Multiscale Simulations

Abstract:

Understanding protein corona formation at the molecular level is essential to various applications such as drug delivery and biomolecule-detection. In this work, we studied ovispirin-1 and lysozyme protein coronas on bare gold and silver nanoparticles by employing hybrid coarse grained and all atom molecular dynamics simulation. Our study shows that protein corona formation is governed by not only protein–surface but protein–protein interactions, as well as the surface hydrophobic effect have significant contributions. The post corona structure was found to be highly dependent on protein types and the size of nanoparticles. The decrease in nanoparticle size leads to more angular degrees of freedom for protein adsorption orientation which also varies based on the nanoparticle in use. Subsequent atomistic molecular dynamics simulations demonstrates the loss of secondary structure of ovispirin after corona formation and the heterogeneity of its local structure.

629830 Mechanistic Understanding of the Biological Responses to Polymeric Nanoparticles

Abstract:

Polymeric nanoparticles have demonstrated tremendous potential for theranostic and nanomedicine applications due their central role in the delivery of a wide range of therapeutic and imaging agents [1-4]. Nonetheless, increasing studies have reported the failures of these nanomaterials in realizing their designed objectives because of a single or combination of factors, notably non-specific biomolecule adsorption, inefficient cellular internalization, triggered immune response, and fast body excretion, which have collectively challenged the merit of nanomedicine for disease theranostics [5-7]. Although numerous efforts have attempted to regulate the biological behaviors of polymeric nanoparticles by controlling their physicochemical properties, a large number of the demonstrated studies have focused solely on either the biological effects induced through the regulation of a single nanoparticle parameter or the separate molecular, cellular, and *in vivo* behaviors of nanoparticles. Crucially, the interdependency of multiple physicochemical properties of polymeric nanoparticles in determining their biological effects from molecular to organism levels is still largely unclear, which warrants a thorough investigation.

In this work, we prepared a series of polymeric nanoparticles consisting of various biocompatible polymeric nanocarriers (i.e., polystyrene (PS), poly(lactic-*co*-glycolic acid) (PLGA), and polyethylene glycol (PEG)) encapsulating organic fluorophores [8]. Subsequent to the synthesis of these nanoparticles, we evaluated their physicochemical properties, such as surface morphology, size distribution, zeta potential, lipophilicity, optical absorbance, and fluorescence. Next, through a range of experimental characterization and molecular simulation, we systematically probed the interdependent effect of nanoparticle lipophilicity, zeta potential, and size on the formation of protein corona around the nanoparticles as well as the *in vitro* cellular uptake and *in vivo* biodistribution of nanoparticles in a zebrafish model.

We observed that nanoparticle lipophilicity influenced the recruitment of non-specific biomolecules, where the adsorption of biomolecules could be reduced by decreasing nanoparticle lipophilicity and vice versa. For a given lipophilicity, the nanoparticle size had a more significant effect in determining biomolecule adsorption than the nanoparticle surface charge. However, irrespective of size, the surface charge of nanoparticles was noted to affect their endothelium and macrophage uptake as well as circulation lifetime. More clearly, the negatively charged nanoparticles could be internalized preferentially by endothelial cells without the need for targeting ligands. Additionally, these nanoparticles were minimally internalized by macrophages and showed a much prolonged *in vivo* circulation lifetime. Based on our experimental data, we proposed a two-step framework to rationally design a single polymeric nanoparticle system with highly regulated *in vitro* and *in vivo* biological behaviors.

In summary, the optimization of the interplay between nanoparticle lipophilicity and surface charge is necessary to control the overall biological behaviors of polymeric nanoparticles. Our study has offered a strong basis for the engineering of enhanced polymeric nanoparticle systems to regulate multiple biological events simultaneously for *in vivo* theranostic and nanomedicine applications.

References

[1] Elsabahy et al., *Chemical Reviews* 115 (2015), 10967-11011.

[2] Ulbrich et al., *Chemical Reviews* 116 (2016), 5338-5431.

[3] Kenry et al., *Advanced Materials* 30 (2018), 1802394.

[4] Kenry et al., *Accounts of Chemical Research* 52 (2019), 3051-3063.

[5] Wilhelm et al., *Nature Review Materials* 1 (2016), 16014.

[6] Rosenblum et al., *Nature Communications* 9 (2018), 1410.

[7] *Nature Nanotechnology* 14 (2019), 1083.

[8] Kenry et al., *ACS Nano* 14 (2020), 4509-4522.

623788 Predicting Protein Adsorption to Engineered Nanoparticles with Supervised Machine Learning Models

Abstract:

Engineered nanoparticles are poised to transform sensing, imaging, and delivery in biological systems.[1,2] In particular, single-walled carbon nanotubes (SWCNTs) are uniquely suited for biological sensing and imaging due to the tissue-transparent and photostable near-infrared fluorescence that they exhibit.[1] As such, SWCNTs have been functionalized with various biomolecules, including synthetic peptides[3] and proteins,[4] to construct optical nanosensors. Optimizing these biomolecule-nanoparticle interactions is key in enhancing nanotechnology function. Yet, functionalized SWCNTs and nanotechnologies more broadly suffer from unpredictable and often unfavorable interactions with the biological environments in which they are applied. When nanoparticles are introduced into biological systems, endogenous proteins rapidly bind and often lead to decreased ability of the nanoparticles to perform their intended functions.[5,6]

In this work, we develop a classifier to predict protein-nanoparticle association. Our purpose is two-fold: as one objective, we aim to predict protein-nanoparticle interactions in full biological environments, informing implementation of appropriate anti-biofouling strategies or exploitation of the adsorbed proteins towards effective biological outcomes. Secondly, we intend to predict high-affinity protein binders to nanoparticles to improve the process of protein-nanoparticle construct design. Toward these ends, we build our classifier based on mass spectrometry-based proteomic data that quantitatively characterizes proteins bound to our SWCNT-based nanosensors.[5] Using widely available protein data, we construct a protein property database and next train and validate a random forest classifier to predict whether proteins adsorb to the nanoparticles. We identify distribution changes among the most important protein properties driving binding. Lastly, we apply the classifier to rapidly identify candidate proteins with high binding affinity to SWCNTs and experimentally validate adsorption of these proteins with a kinetic exchange assay to evaluate the predictive power of our model.[6] In sum, this classifier serves as a valuable method to both overcome the high failure rate in translating nanotechnologies from *in vitro* validation to *in vivo* deployment, and to aid in rational design of future nano-bio tools.

References

1. Hong, G., Antaris, A. L. & Dai, H. Near-infrared fluorophores for biomedical imaging. *Nature Biomedical Engineering* **1**, 1–22 (2017).
2. Mitchell, M. J. *et al.* Engineering precision nanoparticles for drug delivery. *Nature Reviews Drug Discovery* 1–24 (2020) doi:10.1038/s41573-020-0090-8.
3. Chio, L. *et al.* Electrostatic Assemblies of Single-Walled Carbon Nanotubes and Sequence-Tunable Peptoid Polymers Detect a Lectin Protein and Its Target Sugars. *Nano Lett.* **19**, 7563–7572 (2019).
4. Pinals, R. L. *et al.* Rapid SARS-CoV-2 Spike Protein Detection by Carbon Nanotube-Based Near-Infrared Nanosensors. *Nano Lett.* (2021) doi:10.1021/acs.nanolett.1c00118.
5. Pinals, R. L. *et al.* Quantitative Protein Corona Composition and Dynamics on Carbon Nanotubes in Biological Environments. *Angewandte Chemie International Edition* **59**, 23668–23677 (2020).
6. Pinals, R. L., Yang, D., Lui, A., Cao, W. & Landry, M. P. Corona Exchange Dynamics on

Carbon Nanotubes by Multiplexed Fluorescence Monitoring. *J. Am. Chem. Soc.* **142**, 1254–1264 (2020).

622249 DNA-Launched HPV E7 Nanoparticle Vaccine Achieves Long-Term Protection Against Tumor Challenge *In Vivo*

Abstract:

HPV-induced cancers constitute a pressing global public health issue, accounting for over 600,000 cases per year worldwide. By age 50, more than 80% of all individuals have been exposed to HPV infection. Prophylactic vaccines developed in recent years are reducing new HPV infections, but current vaccine formulations fail to induce the cellular immunity necessary to address those with already established HPV infections or malignancies. Nanotechnology-based approaches have provided promising new strategies in anti-cancer immunotherapeutic vaccines through the decoration of multiple immunogens on nanoparticle scaffolds; however, their clinical translation have been hampered by challenges with scalability and purification. To address these challenges, here we present the design and characterization of a novel DNA-encoded nanoparticle vaccine (DNP) scaffolding HPV E7 HLA-Kb epitope with lumazine synthase for inducing anti-tumor immune responses. We demonstrate that through electroporation-facilitated intramuscular DNA vaccination, the scaffolded peptide is produced and self-assembles *in vivo* into 60-mer nanoparticles. Compared to CpG-adjuvanted peptide vaccines, the DNP induced significantly greater antigen-specific CD8+ T cell responses. This translated into tumor protection in 100% of mice when administered prophylactically and complete tumor regression in 50% of mice upon therapeutic treatment in a flank tumor model of HPV-induced cancer. Anti-tumor immunity was retained upon rechallenge over 150 days after initial tumor challenge, supporting the induction of potent immune memory. Finally, in a metastatic lung tumor model, DNP treatment demonstrated prolonged survival and significantly improved lung pathology compared to the CpG-adjuvanted peptide vaccine. These promising results, along with its facile characterization and scalability, support DNA-launched nanovaccines as a promising new platform for inducing potent anti-tumor cellular immunity ideally positioned for study in clinical translation.

632557 Conjugation of CRISPR Machinery to Carbon Nanotubes for DNA-Free Genetic Engineering of Plants

Abstract:

Introduction of CRISPR machinery to a host plant genome may be accomplished through delivery of plasmids encoding Cas9 and sgRNA, *in vitro* transcribed Cas9 mRNA and sgRNA, or recombinant Cas9 ribonucleoprotein (RNP). However, introducing the biomolecular workhorses of CRISPR into model and crop species is hindered by the molecular delivery challenge across the plant cell wall. The cell wall presents a rigid barrier to biomolecule delivery which can be overcome by *Agrobacterium* or biolistic particle bombardment, the success of which is limited by host species, tissue type, and random transgene integration. Notably, transgenic crops are tightly regulated in many countries, wherein these regulations present significant financial and social barriers to the success of transgenic crops. Furthermore, the necessity of tissue culture introduces further limitations, as many crops are recalcitrant towards regeneration. Thus, there exists a great need to engineer model and crop plant species using non-tissue culture based methods and in the absence of recombinant DNA that could introduce risk of host integration.

RNA and RNP delivery simplify the workflow of plant gene editing by circumventing the need for plasmid optimization and improve specificity by reduction of off-target cleavage. Currently, introducing Cas9 RNPs or Cas9 mRNA/sgRNA into plants is not easily addressed by current delivery technologies, as *Agro* is only amenable to DNA delivery and biolistics rely on protein dehydration onto the carrier surface which can result in loss of Cas9 activity.

Nanomaterials offer an addition to the workhorses of plant genetic engineering due to their ability to load diverse cargo, traverse the cell wall and plasma membrane, and selectively localize in tissues and organelles. Prior work has shown that pristine and cationic polymer-functionalized carbon nanotubes (CNTs) act as efficient carriers of siRNA and DNA plasmids, respectively, for the delivery of active biomolecules to plants via aqueous infiltration to the leaf abaxis [1]. We now present the development of DNA-free CRISPR-CNT conjugates for mRNA/sgRNA and RNP delivery to mature plant tissue without biolistics. While RNA delivery proceeds through direct adsorption of active mRNA or sgRNA to the pristine CNT surface, protein delivery necessitates an intermediate ligand to promote Cas9 adsorption to CNTs [2]. We present the design and synthesis of modular peptoid ligands with two domains, a CNT-binding domain and an RNP-binding domain, and demonstrate screening of a peptoid library based on these domains for non-covalent binding of Cas9 to peptoid-CNTs. We discuss the characterization of these diverse conjugates and their potential for use in genome editing applications.

1. Demirer, G. S. et al. *Nat. Nanotechnol.* **2019**, *14* (5), 456–464.

2. Demirer, G. S. et al. *Science Advances* **2020**, *6* (26), eaaz0495.

621781 Extracellular Vesicles As Potentiators of Stress Signals to Alter Placental and Fetal Development

Abstract:

Introduction: Maternal stress is linked to a variety of adverse fetal and pregnancy outcomes. In our well-established mouse model of early prenatal stress (EPS), we observe impaired cognitive function, alterations in metabolic programming, and heightened stress sensitivity, specifically in male but not female offspring. Transcriptomic analysis of the placenta demonstrated that EPS impacts endo- and exosomal processes in the placenta, supporting an involvement of extracellular vesicle signaling. Extracellular vesicles (EVs) are cell-derived nanoparticles that play an important role in intercellular communication, by transporting enzymes, immune signals, and nucleic acids between cells and tissues. During pregnancy, circulating EVs increase in number, in part due to the generation of EVs by the placenta. We hypothesize that, during pregnancy, EVs play an important role in communication between maternal and fetal tissues, and that EV contents and characteristics are altered in response to the maternal environment. Here, we utilize our established EPS murine model to examine circulating maternal EVs over the course of pregnancy, and to probe the mechanism by which maternal EVs impact fetal brain development.

Methods: We first addressed the role of stress in EV signaling by isolating circulating EVs from donor dams in control and EPS-exposed groups. We isolated EVs from plasma collected on embryonic days 12.5, 15.5, and 18.5 (E12.5, E15.5, E18.5) using size exclusion chromatography. Nanoparticle tracking analysis was used to measure the size, concentration, and surface charge of isolated EVs. Next, small RNA-sequencing and proteomics were used to analyze the content of EVs from control and EPS-exposed dams. Labeled EVs were injected into naïve dams, and an *in vivo* imaging system (IVIS) was used to detect EV trafficking in maternal and fetal compartments. A separate cohort was used to detect transcriptomic changes in placental and fetal tissues after exposure to isolated EVs from either control or EPS-exposed dams. EVs were administered to a cohort of naïve dams, and offspring were monitored throughout adulthood for growth and hypothalamic-pituitary-adrenal (HPA) stress axis reactivity.

Results: We observed that, while the size of circulating EVs remains the same over the course of gestation (~120 nm), the concentration of EVs tends to decrease as gestation progresses. While it was not significant, we observed a lower concentration of EVs isolated from E18.5 EPS dams, than from E18.5 control dams. We observed the protein content of EVs isolated from derived from EPS-exposed dams showed a significant decrease in proteins related to the immune response, and an increase in proteins related to metabolic processes. IVIS revealed that injected EVs traffic to the placenta and to the maternal brain, regardless of EPS exposure. RNA-sequencing revealed that control-derived EVs were able to significantly alter protein targeting and messenger RNA catabolic processes in male placentas from EPS-exposed dams.

Conclusions: We have demonstrated the potential for EVs to have a lasting impact on placental and fetal neurological development. Together, these studies provide insight into the role EVs play in promoting stress signals in both maternal and fetal circulation, their interaction at the level of the placenta, and the impact of prenatal stress on important signaling dynamics between maternal and fetal compartments during gestation. Future work will be aimed at engineering biomimetic lipid nanoparticles to target the placenta and fetal brain and rescue the EPS phenotype observed in male offspring.

Funding: NIEHS ES028202, NICHD HD097093, NIMH MH104184 & MH108286

622861 L-Glutamine Delivery to Erythrocytes Via DOPC-DPPG Mixed Liposomes: A Facile System Towards the Treatment of Sickle Cell Disease

Abstract:

Sickle cell disease (SCD) is an erythrocyte-based chronic mortal disease resulting from a point mutation in the β chain gene in the hemoglobin A (HbA) molecule It is known to be one of the most common inherited diseases in the world with a low life expectancy. Although several treatment methods are applied, the pathophysiology of the syndrome is complex and there is still a serious need for development of an effective method for inhibiting or preventing the polymerization and thus sickling of the erythrocytes. Erythrocyte polymerization in SCD is mostly seen as a result of the decrease in the amount of ions and water in the cell, i.e. dehydration and deoxygenation of erythrocytes. Deoxygenated and dehydrated erythrocytes become susceptible to clustering, causing clogging of blood vessels which then leads to crisis. Therefore, development of a treatment method that can effectively prevent deoxygenation of erythrocytes or reduce the oxidative stress of sickle erythrocytes is one of the important issues towards the treatment of SCD. Among a wide variety of potential drug carriers, liposomes are advantageous and preferable with their easy preparation and biocompatibility.

Within this context, in this study we aim to encapsulate L-Glutamine (L-Gln) in liposomes and design a drug carrier system opposed to deoxygenation of erythrocytes in SCD. L-Gln is known to inhibit oxidative stress and its usage is also approved by the US Food and Drug Administration (FDA) to reduce acute pain in SCD patients. To our knowledge, this is the first study that suggests and develops a liposomal drug delivery system for L-Gln administration to erythrocytes and investigates the interactions between red blood cells and drug loaded liposomes. With this scope, the results of this study provide important steps towards obtaining an effective treatment method for SCD.

As a first step of developing drug carriers, we prepared L-Glutamine loaded liposomes composed of 1,2-Dioleoyl-sn-glycero-3-phosphocholine (DOPC) & 1,2-Dioleoyl-sn-glycero-3-phospho-rac-(1-glycerol) sodium salt (DPPG). We have studied the effect of increasing L-Gln content in liposomes by encapsulating L-Gln in three different concentration, namely, 20 mM, 40 mM and 60 mM. In addition, we investigated the physicochemical properties of both unloaded and loaded liposomes as well as analyzing their interactions with red blood cells. Therefore, preliminary results were obtained for further usage of L-Gln encapsulated liposomes towards the reduction and elimination of deoxidation of erythrocytes.

Liposomes were prepared with the well-known thin film hydration method. Afterwards, morphological analysis and physicochemical characterizations of the empty & L-Gln loaded liposomes were performed via several methods such as zeta potential measurements, transmission electron microscopy (TEM), scanning electron microscopy (SEM), differential scanning calorimetry (DSC) and Fourier Transform Infrared Spectroscopy (FTIR). Also, drug encapsulation efficiency and release behavior of the liposomes were characterized via high pressure liquid chromatography (HPLC). Effect of the encapsulated amount of L-Gln was investigated by encapsulating the drug at three different concentrations (i.e.20 mM, 40 mM and 60 mM). Investigation of the interactions between drug loaded liposomes and erythrocytes were implemented with hemolysis experiments and also erythrocytes were visualized via optical microscope and SEM before/after incubating red blood cells with drug loaded liposomes.

The sizes of the liposomes were obtained as 179.9±10.4, 182.3±12.1, 185.0±13.3 and 200.2±15.2 nm for unloaded and for 20 mM, 40 mM, 60 mM of L-Gln loaded liposomes, respectively. PDI values of liposomes were determined to be in range of 0.321 to 0.404, which indicate the homogeneous size distribution of liposomes within the suspension. SEM and TEM images (presented in the Figure (a)-(g)) had also supported the spherical liposome formation and also, sizes of the liposomes in both SEM and TEM images were determined to be in good agreement with the results of size measurements obtained with

Zeta sizer. Zeta potential values of the liposomes were obtained to be in -22.2±1.3 mV and -20.5±1.6 mV range for both unloaded and drug loaded liposomes, which are in good correlation with the results reported for DOPC/DPPG mixed liposomes in the literature. Thermal analysis with DSC thermograms had showed us that the lipids melt cooperatively in the bilayers and they are fully miscible. Moreover, as encapsulated L-Gln amount is increased, melting temperature of the liposomes displayed a shift to lower temperatures, supporting the incorporation of drug within liposome structure had induced a more fluid bilayer. Also, FTIR results indicated the formation of hydrogen bonds between the head groups of lipids and L-Gln. The encapsulation efficiency of L-Gln was determined to be the higher than the ones reported in the literature: 83.6%, 87.1% and 84.9% for 20 mM, 40 mM and 60 mM Gln, respectively. From *in vitro* drug release studies, it was found that after 6 hours, release profile had reached to equilibrium and liposomes loaded with 40 mM and 60 mM of L-Gln had released 42.4% and 45.7% of L-Gln, respectively.

After characterizing the physicochemical properties of the liposomes and release profile of L-Gln, the effect of the presence of L-Gln loaded liposomes on the erythrocytes was also investigated through hemolysis assay and also via the images acquired with optical microscopy and SEM. From the hemolysis experiments, after incubation of drug loaded liposomes (with varying L-Gln concentration, i.e. 20 mM, 40 mM and 60 mM) with erythrocytes at 37 °C for 3 hours, we determined that the presence of liposomes and their interactions with erythrocytes do not cause any significant damage on the structure and morphology of the cells, as this result was also supported with the optical microscopy images of the cells before/after the incubation. Moreover, SEM images of the erythrocytes were acquired and again as supported with the optical microscopy images, it was concluded that the presence of L-Gln loaded liposomes had not only caused no hemolysis, but also, they provide a synergistic effect on the preservation of the structure of red blood cells. We propose that adsorption of liposomes to the membranes of erythrocytes had increased the stability of the cells and the presence of lipids support the erythrocyte membrane to retain its integrity (Figure (h) and (i)). In fact, this is another important result of the study which proves the positive effect of the presence of L-Gln loaded liposomes on erythrocytes.

In conclusion, L-Gln loaded PC/PG liposomes provide preliminary yet promising results in terms of developing a new drug delivery platform towards SCD, for the first time in the literature. Although *in vivo* experiments should also be conducted in the future studies, we believe PC/PG liposomal carriers developed in this study carry a great potential to be used in prevention of deoxidation of erythrocytes which induce polymerization and sickling of cells and lead to painful crisis in SCD.

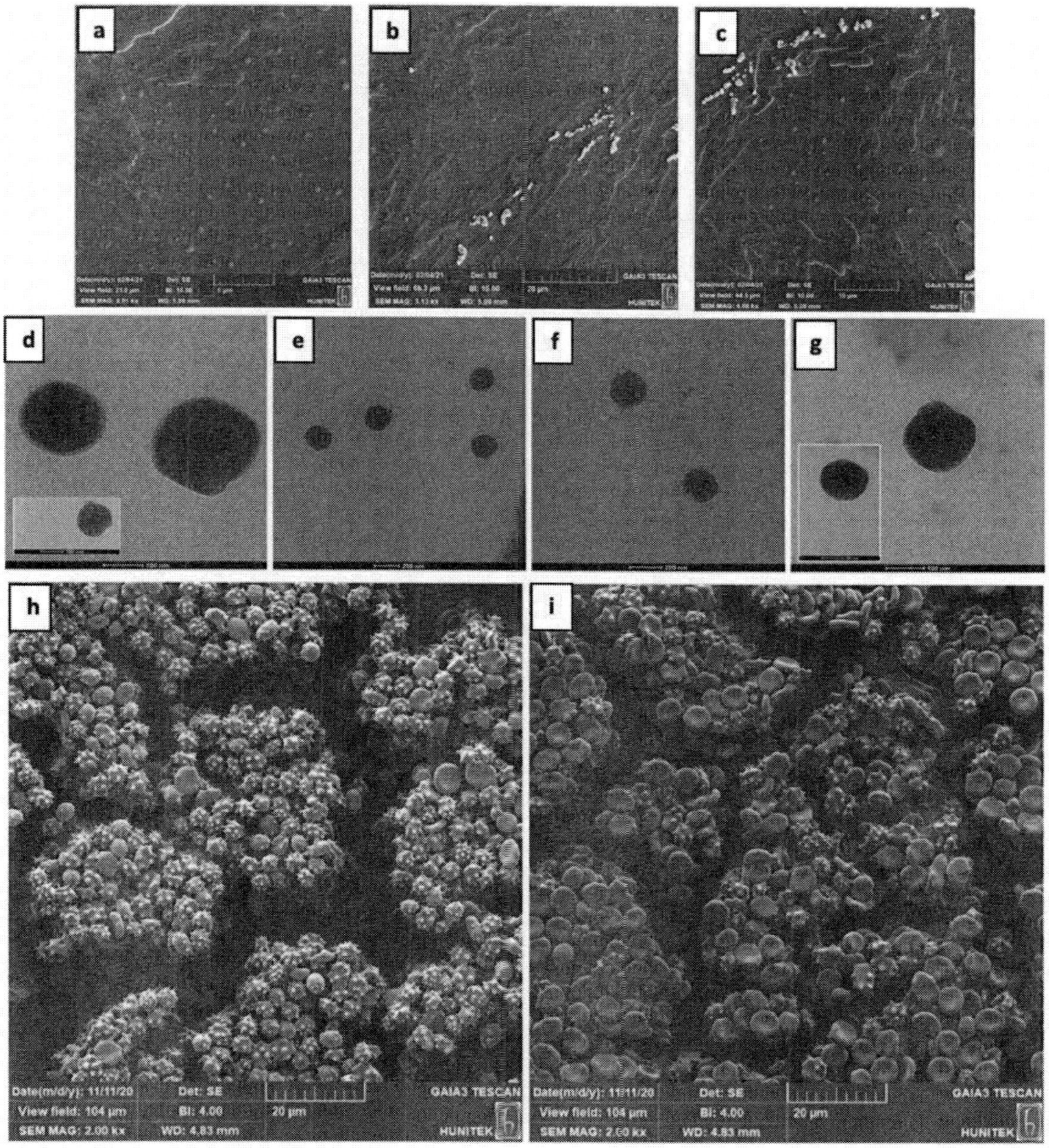

Figure 1. (a), (b), (c) SEM images of liposomal formulations loaded with 20 mM, 40 mM, 60 mM L-Gln, respectively. (d) TEM images of unloaded liposomes, (e), (f), (g) TEM images of liposomal formulations loaded with 20 mM, 40 mM, 60 mM L-Gln, respectively. (h) SEM image of the erythrocytes without the presence of liposomes, (i) SEM image of the erythrocytes that have been interacted with L-Gln loaded liposomes that loaded with 60 mM L-Gln.

630244 Developing Modified Oxalate-Based Doxorubicin (OX-Dox) Prodrugs for Reactive Oxygen Species (ROS) Responsive Drug Delivery

Abstract:

While chemotherapy is a top option for treating metastatic breast cancer, reducing the side effects along with maximizing the therapeutic efficacy remain as major challenges. A prodrug design with a stimulus-responsive linker can address the adverse off-target side effects by deactivating potent but toxic drugs upon conjugation to a linker and reactivating them upon release triggered by a stimulus-response mechanism. Reactive oxygen species (ROS) play a pivotal physiological role in intracellular signaling of any living organism. Due to the elevated levels of ROS in cancerous cells compared to normal cells, an increasing number of ROS-responsive probes and prodrugs have been fabricated in the fight against cancer.

Here, we have developed a modified oxalate-based Doxorubicin (OX-Dox) prodrug by conjugating a benzene ring to one side of the oxalate backbone and a doxorubicin molecule to the other side using an amide bond. Further modification of the functional group on the para position of the benzene ring resulted in a stable but highly specific ROS-responsive OH-OX-Dox prodrug. While the Dox release in PBS was less than 20% within 24 hrs, it was more than 90% after stimulating the production of intracellular ROS by the addition of 50 μM of H_2O_2 and horseradish peroxidase (HRP). An MTS assay result showed that MDA-MB-231 and MDA-MB-468 cells responded to our Dox prodrug in a dose-dependent but less cytotoxic manner compared to free Dox. Also, the addition of L-Buthionine-sulfoximine (BSO), which can enhance the ROS level in cells by inhibiting cellular glutathione (GSH), increased the toxicity of the Dox prodrug towards the aforementioned cell lines that further confirmed the ROS responsiveness of our prodrug. This design has potential to deliver Dox in a sustained manner during systemic circulation at low ROS level while achieving controlled drug release at solid tumor sites at high ROS level.

627156 Real-Time Detection of Salicylic Acid in Living Plants Using Nanosensors

Abstract:

Salicylic acid (SA) is an important plant hormone that regulates numerous plant growth and developmental processes, and activates defences against biotic and abiotic stresses. It is a key signalling molecule that triggers systemic acquired resistance (SAR) in plants, which initiates enhanced resistance in uninfected distal parts of the plant, priming them against further pathogen attacks. Traditional chromatography-based analytical methods of SA quantification are costly and labor-intensive. They are also destructive techniques that could not provide real-time information of SA levels in living plants[1]. More recently, genetically engineered *Acinetobacter sp. ADP1* that produces bioluminescence upon detection of SA have been developed as an *in situ* SA assay[2]. However, application of the biosensor is limited to model plants such as tobacco. Aptamer-based nanosensor have also been studied for rapid salicylic acid detection but applicability of this nanosensor remain limited to SA detection in plant extracts[3]. In this work, we have developed a selective SA plant nanobionic sensor which enables non-destructive, real-time monitoring of SA in living plants. Crucially, the nanosensor will provide new insights into the biological function of SA and its derivatives in SAR and other plant developmental processes. The nanosensor is discovered by screening a library of cationic polymers wrapped around single-walled carbon nanotubes (SWNTs) *in vitro* against a list of common plant hormones analytes. Addition of 100 µM of SA or methyl salicylate (MeSA), the volatile derivative of SA, to the nanosensor resulted in a 35% and 42% near-infrared SWNT fluorescence quenching. In comparison, addition of other plant hormones resulted in <±20% fluorescence change. *Arabidopsis thaliana* mutants in which SA levels are controllable precisely upon inducer treatment have been used to validate and calibrate the SA nanosensor response *in planta*. Further, versatility of the SA nanosensor is demonstrated in non-model *Brassica rapa subsp. chinensis* (Pak Choy) plants, which were infected with *Xanthomonas campestris pv. campestris* (*Xcc*) causing black rot in brassica varieties. In response to *Xcc* infection, SA accumulation is detected by the nanosensor, and confirmed via liquid chromatography–mass spectrometry (LC-MS), in the infected Pak Choy leaves. The SA nanosensor is also interfaced with a portable nanosensor imaging platform, which extends the potential of plant nanobionic approach beyond plant biology studies and towards real-time plant health monitoring in the field.

REFERENCES

1. Matsuura, H.; Aoi, A.; Satou, C.; Nakaya, M.; Masuta, C.; Nabeta, K., Simultaneous UPLC MS/MS analysis of endogenous jasmonic acid, salicylic acid, and their related compounds. *Plant Growth Regulation* **2008,** *57* (3), 293.
2. Huang, W. E.; Huang, L.; Preston, G. M.; Naylor, M.; Carr, J. P.; Li, Y.; Singer, A. C.; Whiteley, A. S.; Wang, H., Quantitative in situ assay of salicylic acid in tobacco leaves using a genetically modified biosensor strain of Acinetobacter sp. ADP1. *The Plant Journal* **2006,** *46* (6), 1073-1083.
3. Chen, C.; Feng, S.; Zhou, M.; Ji, C.; Que, L.; Wang, W., Development of a structure-switching aptamer-based nanosensor for salicylic acid detection. *Biosensors and Bioelectronics* **2019,** *140*, 111342.

626673 Synthesis of Block Carbon Nanotubes and Their Use As Compatibilizers in Immiscible Polymer Blends

Abstract:

A new method to synthesize block nanotubes, i.e. nanotubes with varying chemistry along their lengths, in a scalable manner is described and one example is shown where these nanotubes strengthen polymer-polymer interfaces. The synthetic method relies on the incorporation of nitrogen atoms selectively within the nanotube structure during growth. By rapidly changing the reaction environment, blocks of nitrogen functionality along the length of the nanotube result. Without further modification, we found that the different chemistries did not cause the tubes to partition into different phases in a variety of polymer blends; in the case of interest for this study both tube chemistries would cause the tubes to partition to the poly(methyl methacrylate) phase in a polystyrene/poly(methyl methacrylate) (PS/PMMA) interface blend. Hence, the change in local electronic environment and radical scavenging property of the nanotube near the nitrogen atoms was used to selectively polymerize styrene from the nanotube wall where nitrogen is incorporated. An asymmetric double cantilever beam test was used to determine the ability of these tubes to compatibilize the PS/PMMA interface. Fracture toughnesses increased monotonically with increasing diblock carbon nanotube concentration and maximum values were like those for block copolymer reinforced interfaces while single-chemistry nanotubes showed no reinforcing effect. However, the abrupt increase in fracture toughness with added compatibilizer indicative of a transition to crazing was not found consistent with nanotubes suppressing crazing in homopolymers. Scanning electron microscopy images of the fractured surfaces show agglomerates of carbon nanotubes present which are likely limiting the efficacy of carbon nanotubes at toughening the interface.

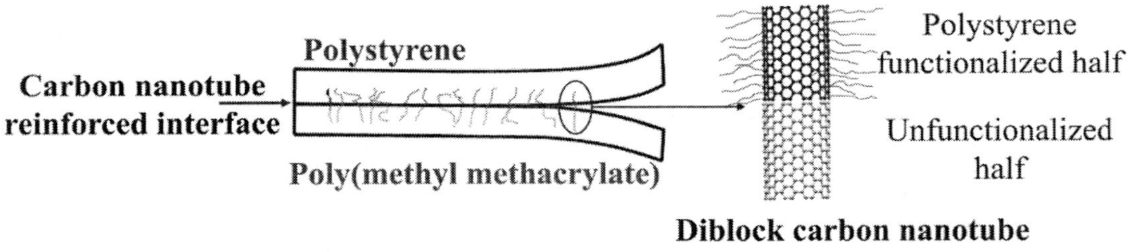

628771 Fundamental Process Parameters and Fiber Formation Mechanisms in Wet-Spinning of Carbon Nanotube Fibers

Abstract:

The mechanical and electrical properties of neat carbon nanotube fiber (CNTF) produced from solutions of pristine CNTs in chlorosulfonic acid are rapidly improving. CNTF is on track to be the strongest synthetic fiber by the end of the decade, and it already surpasses high performance carbon fibers and polymeric fibers in flexibility and electrical/thermal conductivities. As properties begin to exceed those of widely adopted materials (polyaramids, polypropylene, polyethylene, etc.), process cost and environmental efficiency need to be competitive to bring the technology to market. While some research efforts have focused on the impact of the CNT material on CNTF properties, many details of the processing conditions that lead to optimal CNTF properties have not been reported, nor have the mechanisms which limit the formation of CNTF under different conditions. It is understood that CNT solution concentration and spin-line throughput are two primary efficiency determining variables, but the impact of these variables on the spinning window (the range of parameters over which certain grades of CNTF are able to be produced) for the wet spinning process have not yet been thoroughly studied. Here, we seek to define and understand the experimental spinning windows to identify which process parameters govern optimal CNTF properties. We studied the impact of extrusion and take-up parameters, CNT solution concentration, and coagulation conditions on the CNTF properties to determine viable spinning windows. To gain insight and understanding to correlate rheological properties with spinning windows, we measured the viscosity and shear rate behavior of CNT solutions using capillary rheometry. Additionally, we used in-line microscopy on the spinning system to visualize the phenomena occurring at the exit of the spinneret, including die swell, coagulation rate, and mechanisms of filament formation and stability. Finally, we identify the most important factors for CNTF properties and spinning efficiency, and we propose a path forward for efficient production of high-quality CNTF.

629639 Development of Novel Nanoporous Carbon Material for CO_2 Capture

Abstract:

The development of novel materials for efficient CO_2 capture and its exploitation towards sustainable energy is a very imminent approach, due to the glitches of CO_2 emission from diverse industries is directly associated with the issue of global warming. Existing amine absorption technology has several hitches, which are hazardous by products and a high-energy requirement for regeneration, thus research is ongoing worldwide for effective capture of CO_2 by solid adsorbents[1]. In this context, carbon materials are explicitly promising for CO_2 adsorption at ambient conditions, because carbon materials are easily obtainable, economically viable, chemically and hydrothermally robust and also possesses high surface area. Concerning the efficacy of carbon materials for CO_2 capture, we have developed a novel hierarchically structured porous carbon material by templating with hierarchically porous template via CVD method.

Upon successful synthesis of hierarchical nanoporous carbon, an optimized procedure to etch the template is accomplished without any morphological collapse. FE-SEM images in Figure 1 displays the morphology of as-synthesized carbon and template etched carbon materials, demonstrating the same morphology, thus evidencing the preliminary success of this work. The hierarchical nanoporous carbon materials are well characterized by various scientific instrumentation techniques such as XRD, N_2 ads-des measurements, FE-SEM, EDAX, HR-TEM and TGA. Further work on N-doping in the synthesized carbon and its capability towards CO_2 adsorption is currently under progress and all the data will be delivered during the presentation.

Figure 1: SEM images for as-synthesized carbon (A) and template etched carbon (B)

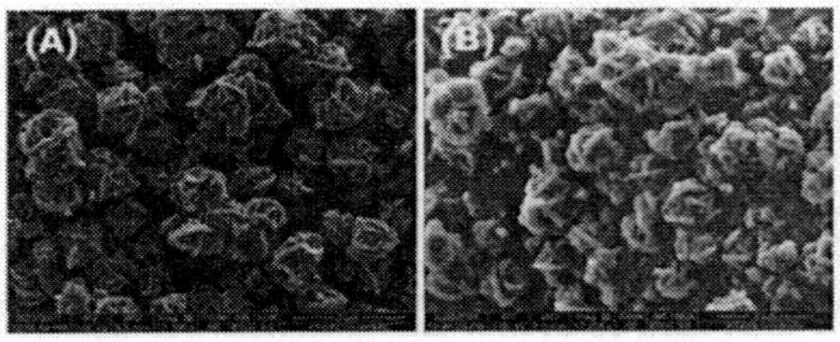

630673 Templated Carbon Nanostructures Synthesis Catalyzed By Group 2 and Lanthanide Oxides

Abstract:

We developed a novel low-cost method for the generation of graphitic carbon with high surface area (> 1000 m^2/g) and high electrical conductivity (3×10^5 S/m). Common minerals or rare earth ions are employed as catalysts, and when combined with water and acetylene large quantities of graphene-like carbon are formed. The carbons have different morphologies depending on the morphology of the catalyst, including crumpled carbon, nanotubes, wrinkled spheres and hollow spheres. The carbons were characterized using powder X-ray diffraction, scanning electron microscopy, transmission electron microscopy, energy-dispersive X-ray spectroscopy, Raman spectroscopy, infrared spectroscopy, simultaneous differential scanning calorimetry / thermogravimetric analysis, multipoint Brauner-Emmett-Teller surface area analysis, quench solid state density functional theory pore size analysis, and X-ray photoelectron spectroscopy. Applications as electrodes for energy storage and for selective adsorption of rare earth ions will be discussed.

631965 Directed Covalent Assembly of Nanodiamonds to Form Continuous Films

Abstract:

Polycrystalline diamond thin films are composed of a mixture of sp3- and sp2-hybridized carbon atoms, inheriting some of the superlative properties from its single crystalline form, such as increased chemical stability with low coefficient of thermal expansion, good thermal conductivity, wide optical transparency, and biocompatibility. These films are most commonly grown via chemical vapor deposition. However the growth process is energy intensive, and requires specialized growth chamber, vacuum conditions, precursor gases, and large power requirement. Further, the growth is restricted to substrates that can resist melting, reaction with process gases, and carbon dissolution at high temperatures. Transfer printing of CVD diamond films onto flexible substrates has been demonstrated; however, this process is limited to feature sizes smaller than 1 mm and the resulting films are prone to breakage upon flexing of the substrate. In this talk, we present a room temperature assembly process for the cyclic attachment of carboxylated ND aggregates (ND-COOH) and a diamine using a carbodiimide cross-linker. The process begins with an amine functionalized surface. In most cases, this can be achieved by hydroxylation with oxygen plasma, followed by silanization with 3-aminopropyltriethoxysilane. The primary amines on the substrate were then reacted with an o-acylisourea active ester form of ND-COOH. This form was realized using 1-ethyl-3-(3-dimethylaminopropyl) carbodiimide hydrochloride. Because the half-life of the acylisourea active ester is close to an hour at 20 °C, the reaction was limited to 30 min. The unreacted active ester groups on the substrate-bound ND-COOH were then quenched with a diamine linker, resulting in an amine-terminated surface. Here, we chose a short-chain length linker, ethylene diamine, to reduce the spacing between ND aggregates. A continous film was achieved by cyclic exposure to freshly activated form of ND-COOH and ethylene diamine.

The morphology of the ND films was examined using atomic form microscopy (AFM) and scanning electron microscopy (SEM), their chemical nature suing X-ray photoelectron spectroscopy (XPS) and Raman spectroscopy, and their thermal transport properties via microfabricated test devices. The effect of solvent media and its pH was studied using deionized water, 10 mM KCl solution, and MES buffers with pH 5.5-7. With deionized water or 1 mM KCl as the reaction medium, when the carbodiimide cross-linker was added to a horn-sonicated ND-COOH suspension in deionized water, its hydrochloride content lowers the pH and leads to particle agglomeration. Further, the particle size distribution in DI water varied greatly. The SEM images of nanodiamond films assembled using DI-water based sol prepared by horn sonication showed higher surface coverage and increased density compared to the films formed by bath sonication.

As seen via SEM and AFM, a pH 6.5 or pH 7 buffer leads to a continuous surface coverage and a similar apparent porosity (~30%), while a pH 6 buffer leads to discontinuous films and more porous films (~65%). Further, the films deposited at pH 7 showed smaller pore sizes and a higher thermal conductivity in comparison to films deposited at pH 6.5. The as-deposited films at pH 7 showed a thermal conductivity as high as 12 ± 2.5 W m^{-1} K^{-1} at 310 K, which is comparable to that obtained for ultrananocrystalline diamond films obtained via chemical vapor deposition. However, sequential thermal annealing of the nanodiamond films at temperature up to 400 °C led to the aggregation of nanodiamond to segregated islands, loss of surface coverage, and increase in porosity. The thermal conductivity of all the annealed samples was comparable near room temperature regardless of deposition pH. The pH 6.5 and pH 7 samples exhibited a general increase in thermal conductivity with increasing temperature, while pH 6 samples exhibited statistically similar thermal conductivity with increasing temperature. The use of a phonon hopping model to deduce the phonon transfer at the grain-grain boundary indicates that annealing works to homogenize interfaces within the films and reduce sample-to-sample variation but it comes at the expense of less efficient phonon transmission across interfaces. Overall, the results demonstrate a way to achieving porous, low-cost nanocrystalline diamond thin films with tunable film morphology and thermal conductivity for electronics and biomedical applications.

632042 Atmospheric Pressure Plasma Nanoparticle Generator for Scalable CNT Production in a Fccvd Process

Abstract:

Carbon nanotubes (CNTs) are one of the most promising candidates for next-generation materials for applications ranging from high-strength, light-weight construction materials to field emission transistors. Unfortunately, wide-scale adoption of CNTs is impeded by the high production cost of crystalline, high aspect ratio CNTs needed for high performance in applications. Floating catalyst chemical vapor deposition (FCCVD) is the most promising synthesis method for scale-up due to its continuous nature and ability to directly produce powders, fibers, films, or foams. However, the FCCVD process currently exhibits low process efficiency and poor reactant utilization leading to production costs that are prohibitively high for most applications.

One major source of process inefficiency is from the catalyst nanoparticles used to grow CNTs. CNT diameter and number of walls is known to be proportional the catalyst nanoparticle diameter. To grow high-quality CNTs with few walls, catalyst nanoparticles need to be below 2 nm in diameter. Larger catalyst nanoparticles (diameters greater than 5-10 nm) will not be able to efficiently grow the desired CNTs. Catalyst utilization, defined as the proportion of catalyst material that successfully grows a CNT, is below 1% in current reactors. This leads to low productivity, low feedstock conversion, and additional post-processing steps to remove residual catalyst. Introducing the catalyst is generally done by the injection of metal precursors, such as ferrocene or iron pentacrabonyl, which decompose in-flight. This process introduces a high degree of complexity to the reaction system as the catalyst formation and CNT growth are occurring simultaneously and are poorly controlled since process parameters are optimized for CNT growth, not catalyst formation. Additionally, the metal precursors contain carbon that is released during thermal decomposition leading to a second carbon source, introducing potential competitive reaction pathways.

In this work, we describe an atmospheric pressure, radio-frequency plasma catalyst nanoparticle generator used to produce a scalable, consistent, and controllable stream of catalyst nanoparticles with mass flowrates that can enable high CNT productivity. Designing a preformed catalyst generator decouples the catalyst formation and CNT growth, allowing for process parameters to be optimized for both separately. A process schematic can be seen in Figure 1. A mixture of argon and hydrogen is used as a plasma gas, and methane is used as a carbon feedstock (in the dilution stream). This process utilizes an elemental iron source, in the form of a wire inserted into the plasma, for the catalyst particles, thus removing extraneous carbon from reaction management. The process geometry also plays a key role, in particular the length of the plasma and growth region before dilution. We present our results in optimizing this process to improve productivity, catalyst utilization, and reactant conversion to few walled, high aspect ratio CNTs.

This work was supported by the US Department of Energy award DE-AR0001015 (Advanced Research Projects Agency - Energy)

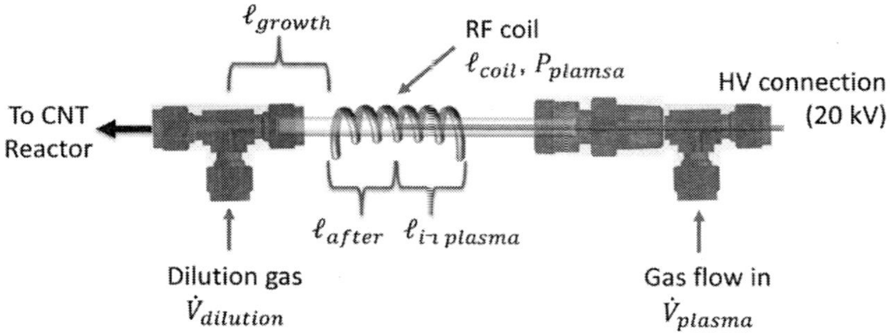

Figure 1. Schematic of atmospheric pressure flow through plasma with important variables.

626924 Impact of Surfactant Chemistry on Single-Wall Carbon Nanotube Extraction Conditions in Aqueous Two-Polymer Phase Extraction

Abstract:

Aqueous two-polymer phase extraction (ATPE) is an effective method for isolating individual (*n,m*) species of single-wall carbon nanotubes (SWCNTs) from multiple specie mixtures, with control of individual SWCNT specie isolation usually achieved by competing two or more surfactants against the nanotube surface. Here, we investigate the effects of systematic variations in surfactant chemistry on ATPE separation conditions using a recently reported near-infrared fluorescence emission monitoring method. Specifically, variations in alkyl surfactant tail length and anionic head group chemistry were observed to strongly affect the amount of surfactant necessary to cause extraction of nanotube species in ATPE extraction. Substitution of different di-hydroxy bile salts is also shown to greatly affect the surfactant concentrations necessary for (*n,m*) extraction. These results provide additional insight into the underlying mechanisms behind ATPE-based SWCNT separations, towards further development and optimization of the ATPE method.

621206 Conversion of Natural Gas to Value-Added Carbon Materials Via Microwave Plasma Technology

Abstract:

Advanced microwave plasma technology practically and cost-effectively converts natural gas to value-added chemicals and premium carbon materials such as graphene and conductive carbon black analogues (CCBA) with no CO_2 emissions and low capital and infrastructure expenditures. In this work a microwave driven plasma drives hydrocarbon decomposition – producing a variety of carbon nanostructures without the use of catalyst. Primary products include nanographene and graphitic carbon particles with structure analogous to conductive carbon blacks. Analytical techniques including high-resolution transmission electron microscopy (HRTEM), X-ray diffraction (XRD), thermogravimetric analysis (TGA), Raman spectroscopy and electrical conductivity measurements are utilized to study the form and quality of these valued carbon materials. With lower-energy requirements than conventional thermal plasmas, reactions in microwave plasmas are driven by electron kinetics rather than thermodynamics, and their non-equilibrium energy distribution opens reaction pathways that are unavailable with conventional chemical or thermal plasma processes.

Whereas pyrolytic decomposition of hydrocarbons typically produces unstructured carbon, the nanoplatelets are characterized by dimensions of 200 – 500 nm, occurring in stack of 2-6 with morphology resembling mildly crumpled paper. Methane conversion can reach 90% with selectivity to products determined by the effluent stream's temperature regime. Accompanying spectral diagnostics are used to characterize the reacting flow. The complement CCBA material has structure analogous to that of commercial high-value conductive carbon blacks, with high electrical conductivity higher than that of some commercially available conductive carbon blacks. These results highlight the importance of advanced plasma technology for the economic utilization of natural gas by producing premium carbon materials.

The co-production of hydrogen and high-value carbon materials from natural gas offers opportunities to reduce the costs associated with large-scale production of industrial or fuel hydrogen or platform chemicals while also producing synthetic, high-value carbon products such as the graphene nanoplatelets. H Quest Vanguard Inc. is developing microwave plasma processing of natural gas as a modular, compact and portable technology for production of hydrogen and premium carbon products.

Figure 1 Nanocarbons formed in the microwave plasma.

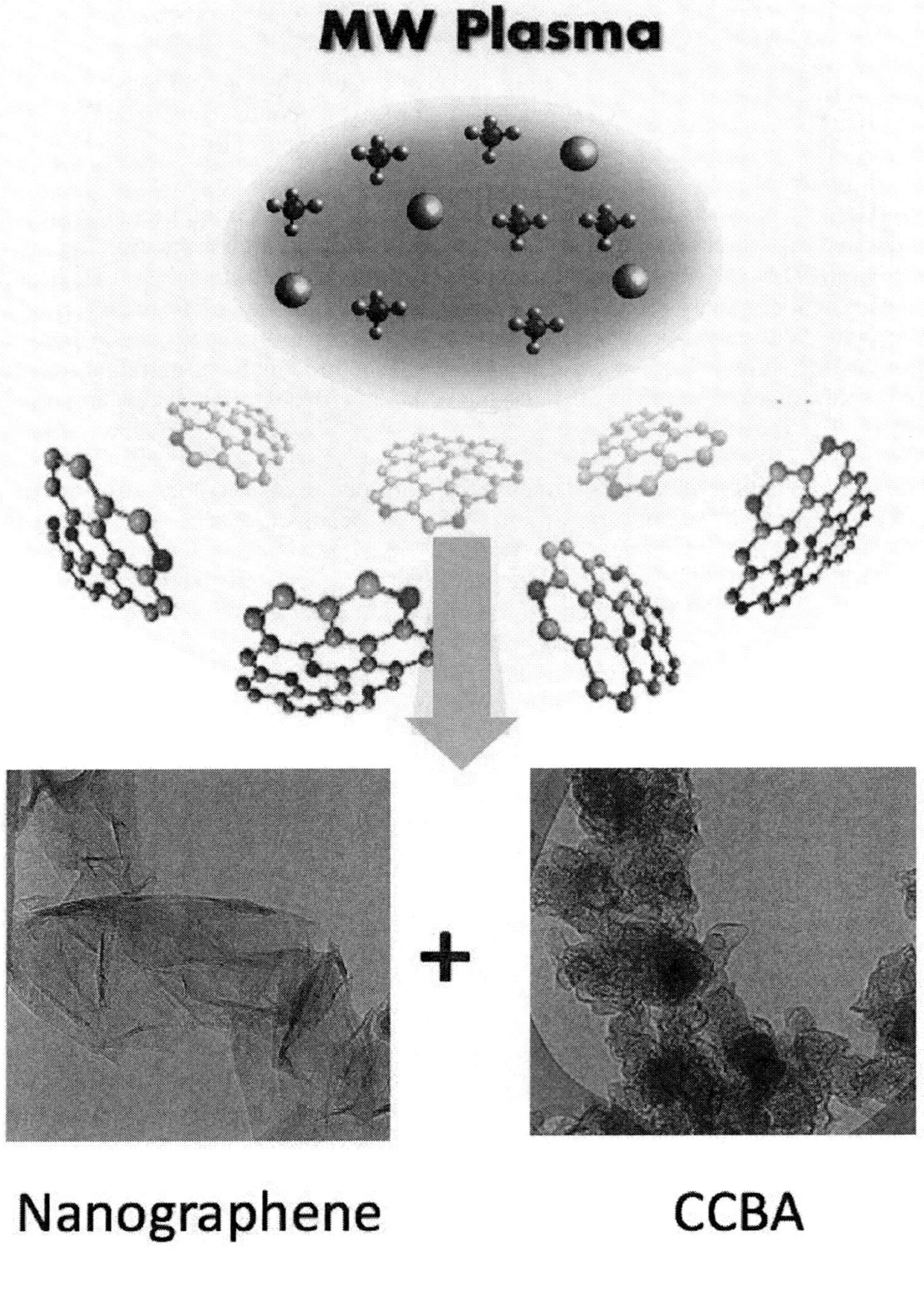
MW Plasma
Nanographene
+
CCBA

624106 Stimuli Responsive Self-Folding 3D Graphene Architectures

Abstract:

Due to their exceptional properties and ultra-thin dimensions, two-dimensional layered materials (2DLMs) have been widely utilized in electronics, optics and drug delivery. Prior research has focused on integration with rigid planar substrates which can limit applicability. Here, we describe monolayer graphene integration with a flexible and gradient crosslinked epoxy polymer SU8. We demonstrate reversible solvent responsive self-folding and unfolding of integrated photodetector devices using these flexible substrates. Experiments are guided by finite element mechanics and electromagnetic simulations with good agreement. We anticipate that this approach could be used to create novel 3D devices using graphene and other 2DLMs of relevance to wearables, flexible electronics, photonics and biosensors.

629841 Fluorescent Single-Walled Carbon Nanotubes for Dilute Divalent Metal Cation Sensing

Abstract:

Colloidal single-walled carbon nanotubes (SWCNTs) offer a promising platform for the nanoscale engineering of molecular recognition. Optical sensors have been recently designed through the modification of non-covalent corona phases of SWCNTs. Specifically, DNA are of great interest due to the breadth of the design space and our ability to control these molecules with sequence specificity at scale. Utilizing these constructs for metal ion sensing is a natural extension of this technology due to DNA's well-known coordination chemistry. Additionally, understanding metal ion interactions of these constructs allows for improved sensor design for use in complex aqueous environments. In this work, we study the interactions between dilute divalent metal cations and DNA-SWCNT under the most controlled experimental conditions for SWCNT optical sensor studies to date. We found that best practices for the study of colloidal SWCNT analyte responses involve mitigating the effects of ionic strength, dilution kinetics, laser power and analyte response kinetics. We also discover that DNA-SWCNTs offer two unique sensing states at pH 6 and 8. The combined set of sensors in this work allows for the differentiation of Hg^{2+}, Pb^{2+}, Cr^{2+} and Mn^{2+}. Finally, we implemented Hg^{2+} sensing in the context of portable detection within fish tissue extract, demonstrating nanomolar level detection.

631744 Investigating the Mechanism of Temperature-Induced Graphene Self-Folding in Water

Abstract:

The popularity of graphene origami - eyed for applications in electronics, energy harvesting, molecular storage and membrane development - has been on the rise for the last few years, with the world's tiniest microchips created through the same process just in 2021. With hopes of making graphene origami easier, temperature-induced self-folding of graphene using thermosensitive polymers like poly(N-isopropylacrylamide) (PNIPAM) has been demonstrated experimentally. This process, attributed to the coil-to-globule transition of PNIPAM, is however, poorly understood, making the creation and control of these self-folded three dimensional (3D) structures difficult. For this reason, we employ coarse-grained (CG) molecular dynamics (MD) simulations to develop an accurate molecular-level mechanism for the self-folding of PNIPAM-grafted graphene in presence of an explicit solvent. Our analyses of contacts, energetics and structure of PNIPAM-solvent interface indicate that the process is actually initiated by the weakening of PNIPAM-water interactions at higher temperatures, resulting in strong hydrophobic graphene-PNIPAM interactions. Densely grafted PNIPAM chains above LCST are observed to stabilize through chain aggregation and a coil-to-globule transition may not always be observed. By further varying system parameters like graphene width and polymer coverage, we show that graphene folding only occurs when the total energy of the final conformation is lower than the total energy of the initial conformation, indicating the stabilization of the structure after folding. We also find that PNIPAM-PNIPAM interactions compete with graphene-PNIPAM interactions and folding occurs when the latter dominates. Through the molecular-level insights obtained in this study, we thus provide energy criteria for graphene-folding that can be essential for controlling and tuning this process *via* temperature control.

632294 Rapid Feedback Experimentation for Floating Catalyst Carbon Nanotube Growth

Abstract:

Carbon nanotubes (CNTs) are among the most promising candidates for the creation of next generation engineering materials, as well as innumerable applications across diverse fields such as medicine and communications. Moreover, they provide a pathway to fix carbon from fossil or biological sources into long-term, useful solid materials. Adoption into large scale application such as composites or cabling, however, is still stifled by the limited synthesis scale of long, highly crystalline few walled CNTs (FWCNTs) needed to attain strength and conductivity at in macroscale applications. Despite major cost-reduction over the past decade, continuous production techniques such as floating catalyst chemical vapor deposition (FCCVD) have not yet attained the requisite the efficiencies to allow economic process scale-up. This is partially due to the large parameter space of the FCCVD process, which includes diverse carbon sources, catalyst chemistries, and process configurations. This makes systematic optimization challenging due to the large number of experimental combinations possible and extremely labor-intensive nature of such work.

To address these issues, the autonomous research system (ARES) was developed at the Air Force Research Laboratory and has proved invaluable by providing rapid feedback experimentation coupled with autonomous or semi-autonomous operation. ARES dramatically decreases the experimental effort required for each condition tested, allowing for both rapid, broad exploration of the CNT growth parameter space as well as repetition for the collection of meaningful statistics on process variability.

Thus far, ARES has been used to optimize surface-bound CTN growth. Here, we apply the ARES to FCCVD growth (FC-ARES) with the goal to develop an autonomous, closed loop research system for FCCVD reactors. Our custom built FCCVD reactor has an inline Raman spectrometer enabling rapid iterations through the FCCVD parameter space with near instantaneous feedback on the crystallinity and composition of the produced CNTs. We present two test cases to demonstrate the utility of this system. The first is a set of 179 experiments exploring how differing molecular structures of related carbon sources (ethanol, 1-propanol, and 2-propanol) impact the morphology of produced CNTs while controlling for flow rates as well as various interatomic ratios. The second is a set of 360 experiments which examines the effect of sulfur to iron ratios on the produced CNTs. It also examines the secondary effect of the oxidative or reducing nature of the system, as controlled by the addition of CO_2 and H_2 respectively. We expect FC-ARES to interrogate the large parameter space and identify impactful parameters that increase process efficiency, yield, and improved control over morphology and crystallinity, thereby defining pathways to geometric scaling of FCCVD reactors.

Figure Caption:

CNTs grown on the FC-ARES show diameter differences depending on the carbon source, as determined from the RBM region's spectral centroid.

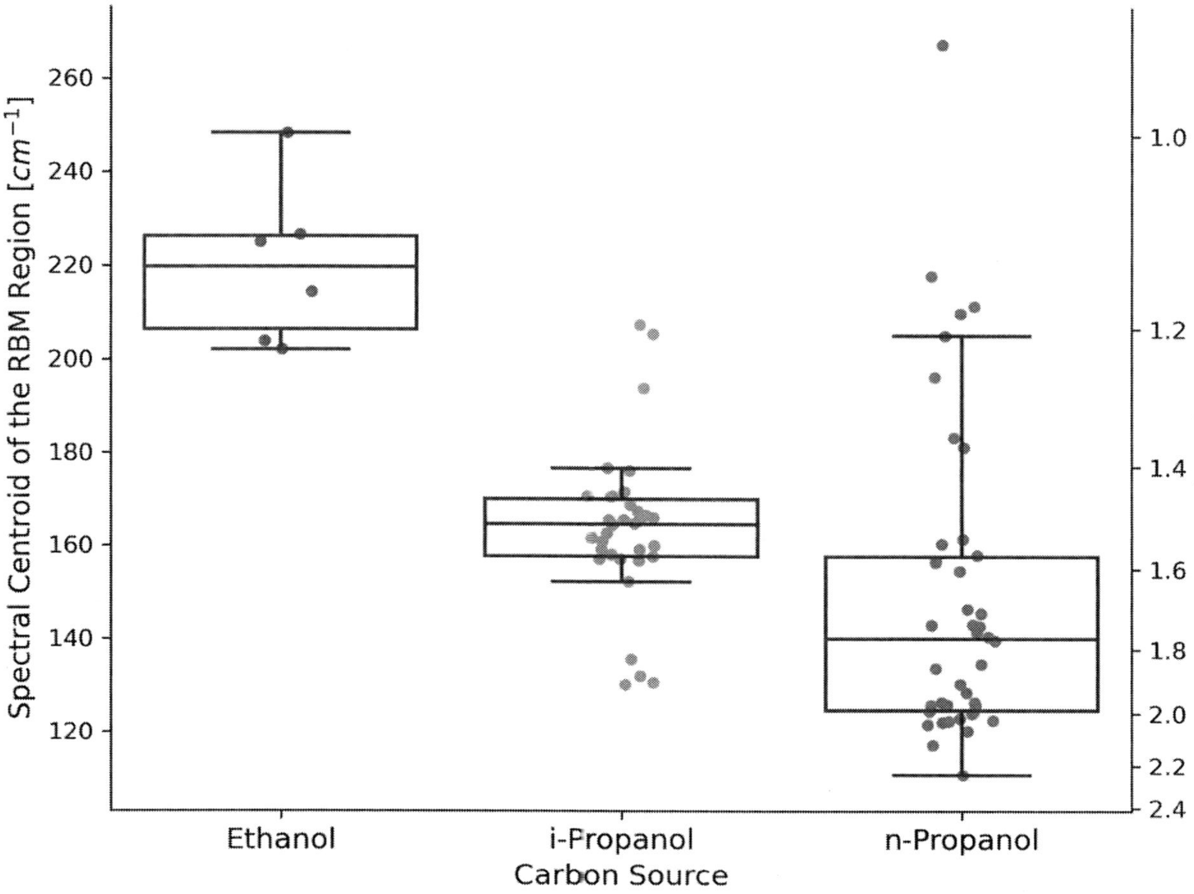
Spectral Centroid of the RBM Region [cm^{-1}]
260
240
220
200
180
160
140
120
Nanotube Diameter [nm]
1.0
1.2
1.4
1.6
1.8
2.0
2.2
2.4
Ethanol
i-Propanol
n-Propanol
Carbon Source

624621 Significant Enhancement of the Hydrophilicity and Ageing Behavior of Coc-Based Microfluidic Devices By Graphene Oxide Thin Layer Deposition Surface Modification

Abstract:

Several materials are used in the fabrication of microfluidic devices. Recently, the fabrication of these devices has relied on polymer-based material such as cyclic olefin copolymer (COC) [1-3]. This attention is related to its characteristics that make fabrication faster, easier, and cheaper through the use of soft lithography in comparison to other materials such as silicon and glass devices [1, 2]. Furthermore, COC has other desirable features such as being biologically inert, optically transparent, non-flammable, low-cost materials and non-toxic[2]. Therefore, it becomes an alternative material for different micro-manufacturing models. However, COC has some limitations, which limit its full application. Unlike silicon or glass substrates, COC is hydrophobic, which makes it not suitable for representing hydrophilic microchannels[1-3]. The surface of the microchannel can be modified to be hydrophilic to ensure that the channel walls are completely wetted so that it can be used for selective applications.

Graphene oxide is one of the most interesting materials in research today due to its superior properties. In this study, different concentrations of graphene oxide (GO) were used to improve the hydrophilicity of the COC surface, thereby enhancing the wettability of its patterned microchannels. Argon-oxygen plasma was used to improve the adhesion between GO and COC surface. Plasma power of 30W with exposure times from 5min to 2hrs were carried out.

A comprehensive study on the influence of plasma exposure time, GO and ageing on the surface properties of COC surface was carried out using different characterization techniques including Scan electron microscopy (SEM), Atomic force microscopy (AFM) and Fourier transform infrared spectroscopy (FTIR). The hydrophilicity of COC was evaluated by measuring the water contact angle of the COC surface with and without GO deposition. As the GO concentration increased, a decrease in contact angle was observed, which confirmed the ability to enhance the wettability of the COC surface. The results also show that, compared with COC treated with plasma alone, COC coated with different concentrations of GO can effectively reduce the aging effect.

References:

[1] W. A. Bauer, M. Fischlechner, C. Abell, and W. T. Huck, "Hydrophilic PDMS microchannels for high-throughput formation of oil-in-water microdroplets and water-in-oil-in-water double emulsions," *Lab Chip,* vol. 10, no. 14, pp. 1814-9, Jul 21 2010, doi: 10.1039/c004046k.

[2] S. Hsu, Z.-Y. Zhang, and C.-W. Tsao, "Thermoplastic Micromodel Investigation of Two-Phase Flows in a Fractured Porous Medium," *Micromachines,* vol. 8, p. 38, 01/26 2017, doi: 10.3390/mi8020038.

[3] A. Alazzam and N. Alamoodi, "Microfluidic Devices with Patterned Wettability Using Graphene Oxide for Continuous Liquid–Liquid Two-Phase Separation," *ACS Applied Nano Materials,* vol. 3, no. 4, pp. 3471-3477, 2020/04/24 2020, doi: 10.1021/acsanm.0c00200.

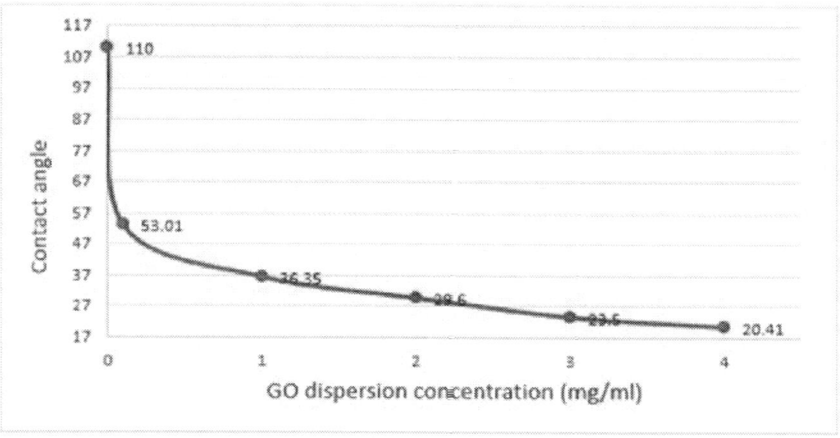

Figure 1. Contact angle of water droplet on 5min plasma pre-treated COC substrate coated with different concentration of graphene oxide (GO).

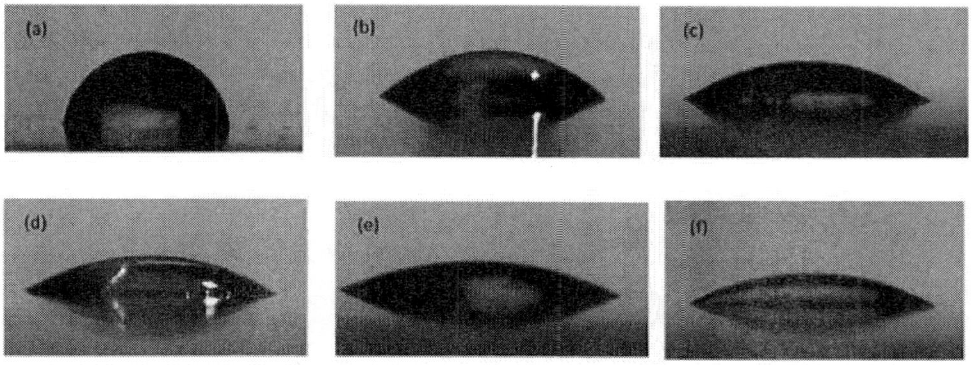

***Figure 2**. Images for water droplets on top of (a) COC surface, (b) GO-coated COC surface (1mg/ml GO), (c) GO-coated COC surface (2mg/ml GO), (d) GO-coated COC surface (3mg/ml GO), (e) GO-coated COC surface (4mg/ml GO), (f) GO-coated COC surface (0mg/ml GO) at 5min plasma exposure time. (Results are collected immediately after surface modification)*

624502 Harmonic Enhancement of Activity and Stability of Iridium-Based Catalysts for Proton Exchange Membrane Water Electrolysis

Abstract:

Recently, the proton exchange membrane water electrolysis (PEMWE) which can store the surplus electricity as a form of transportable hydrogen has been attracting great attention for expanding the large-scale renewable electricity using sun and wind for the carbon-neutral society [1]. Due to the slow oxygen evolution reaction (OER) that can generate the oxygen in the PEMWE and harsh reaction conditions like low pH and high voltage in the anode during the electrolysis, the precious metal oxide catalysts such as IrO_2 and RuO_2 should be adapted in the membrane electrode assembly (MEA). However, there is a challenge to decrease the amount of the noble metal to the level of 0.01 g/kW [2] in the MEA to commercialize PEMWE at a GW per year scale in near future. Thus, the catalysts for the OER, which are mainly iridium(Ir)-based composition, have been developing for harmonic improving the activity and stability through many measures such as mixed metal oxides [3], core-shell nanostructure [4], and supported catalysts [5].

In this presentation, new synthetic approaches for harmonic enhancement of the activity and stability will be discussed. For example, a novel solution process using ethylene glycol, formic acid, and sodium acetate for breaking the trade-off relationship of the activity and stability [6] is presented, which is very suitable for mass production of IrOx catalyst because it is simple and less time-consuming. As the second example, the modified Adams fusion synthesis using glycol as an additive, which can produce a higher surface area of IrOx catalyst is displayed. The harmonized activity and stability obtained by a modified Adams method results in the enhancement of the MEA performance under PEMWE conditions.

[1] IRENA "Innovation Landscape Brief: Renewable Power-to-Hydrogen", International Renewable Energy Agency, Abu Dhabi, 2019 at www.irena.org/publications.

[2] M. Bernt, A. Siebel, H. A. Gasteiger, *J. Electrochem. Soc.,* **165** (2018) F305.

[3] S.W. Lee, C. Baik, T.-Y. Kim, C. Pak, *Catal. Today*, **352** (2020) 39.

[4] C. Van Pham, M. Bühler, J. Knöppel, M. Bierling, D. Seeberger, D. Escalera-López, K.J. Mayrhofer, S. Cherevko, S. Thiele, *Appl. Catal. B.*, **269** (2020) 118762.

[5] S. Zhao, A. Stocks, B. Rasimick, K. More, H. Xu, J. Electrochem. Soc., 165 (2018) F82.

[6] S. W. Lee, C. Baik, D.-H. Kim, C. Pak, J. Power Sources, 493 (2021) 229689.

624549 Photocatalytic Core-Shell Nanotube Array Converting Carbon Dioxide and Water to Fuels

Abstract:

Coupling H2O oxidation with CO2 reduction under membrane separation at the nanoscale is a critical design feature of natural photosynthesis for minimizing efficiency-degrading processes, in particular ion transfer resistance losses and unwanted reactions. To incorporate the feature into an artificial photosystem, we have developed square inch sized arrays of Co3O4-SiO2 core-shell nanotubes. As Figure 1 shows, each tube is able to accomplish the complete photosynthetic cycles, while spatially separating the incompatible oxidation and reduction environments on all length scales from nano to macro.

In-situ monitoring by electrochemical, optical, infrared, and mass spectroscopy plays a central role to demonstrate and optimize the photocatalytic performance of the nanotube array. To be more specific, electrochemical and optical spectroscopy establishes charge-carriers transfer mechanism through the nanotube wall, in-situ infrared spectroscopy illuminates how Ag nanoclusters photo-deposited outside the tubes serve as catalytic sites for visible light-driven CO2 reduction to CO, and mass spectroscopy allows to quantify O2 evolution from visible light-driven water oxidation inside the tubes.

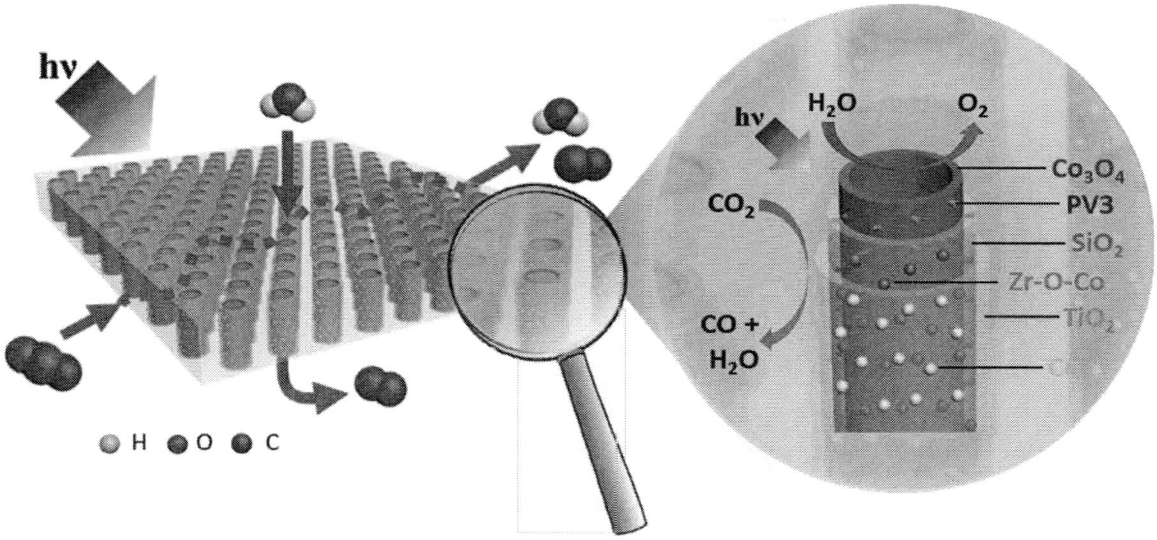

624640 Impact of Crystalline Structure on Photodynamics in Halide Perovskites

Abstract:

Metal-halide perovskite (MHP) is a promising material for the next-generation solar cells and/or light emitting diodes (LEDs) due to its outstanding photophysical properties. Of particular, the perovskite structure consists of corner-shared octahedra, of which rotation (i.e., octahedral tilting) yields various polymorphic phases. For example, the $CsPbBr_3$, which is one of the most promising materials for the LED application, has orthorhombic, tetragonal, and cubic phases, and the polymorphic phase is known to be controlled via nanostructuring. In this study, we investigate the impact of crystalline structure on photodynamics in the MHPs. Using time-domain *ab-initio* molecular dynamics simulation, we compare the carrier dynamics in different structural phases of $CsPbBr_3$, and unravel how the lattice vibration of the different structural phase-changes the carrier lifetime. Throughout our current theoretical study, we anticipate providing a mechanistic guideline to enhance the quantum yield of the MHPs via nanostructuring, which will offer a stepping stone towards a successful deployment of MHPs for optoelectronic applications.

625288 Phase Transition and Metal Exsolution in Perovskites to Enhance the Catalytic Activity for Hydrogen Production and Conversion

Abstract:

Redox exsolution, *in-situ* synthesis process of metal nanoparticles upon high-temperature reduction has widely been studied as a way to fabricate metal-decorated oxide catalysts in the fields of high-temperature catalysis and renewable energy. In perovskite oxides, the exsolution often accompanies the phase transition, which in turn increases the catalytic activity. In this talk, we will discuss about our exploration of the possibility and detailed mechanism for metal exsolution with phase transition on various kinds of perovskite oxides. Using our combined experimental and theoretical approaches, we also demonstrate that these catalysts can be applied to high performance high-temperature water-gas shift reaction and solid oxide fuel cell anode reaction.

625309 First-Principles Design on Small Cluster Catalysts for the Electrochemical NH_3 Synthesis By Ligand Engineering

Abstract:

The ammonia production via electrochemical nitrogen reduction is one of the most attractive and emerging chemical process in hydrogen storage applications. Especially, the development of highly efficient catalyst is essential for the electrochemical ammonia synthesis process. However, such development is still challenging issue owing to the difficulty of direct characterization of electro-catalyst property. Quantum mechanics-based first-principles density functional theory (DFT) calculation is one of flexible and powerful means which can provide the quantitative information on the electro-catalysis and in turn help to design the novel electro-catalyst composition. In this presentation, we examined the electrochemical nitrogen reduction reaction (N2RR) on the small cluster metal catalysts embedded in support using DFT calculation. We chose the single, double and triple atoms for the size of small clusters and also investigated the electrochemical N2RR catalysis for small multimetallic clusters through ligand engineering. For the support of small metal cluster catalysts, two dimensional materials such defective graphene and mexene are considered. In addition, the electronic structure was analyzed for the clear understanding of activity of small clusters toward ammonia production via electrochemical N_2 reduction. Our theoretical study provides the fundamental mechanism of NH_3 production catalysis on small clusters and gives the physical and chemical intuition for the next generation bimetallic small cluster catalysts for hydrogen storage applications.

630448 A General Approach to Synthesize Free-Standing Metal Selenides@ Carbon Nanofibers Anode for Lithium/Sodium Ion Batteries

Abstract:

The trend of developing lighter, smaller, long-lasting portable devices has driven researchers to explore the next generation materials for lithium ion batteries (LIBs) and beyond including sodium ion batteries (SIBs). Transition metal selenides are a notable group that could not only theoretically provide higher specific capacities than commercial graphite anode through conversion reaction, but also show good conductivity, stability, as well as cost-effectiveness. Herein, we report a general approach to develop free-standing metal selenides/carbon nanofibers (CNFs) composites as the anode for LIBs/SIBs via electrospinning. The structure-property-performance relationship of the composites of different metal selenides (MSe, M=Sn, Fe, Ni, Cu) anchored on the CNFs was thoroughly investigated in LIBs system. The potential application in SIBs was also explored. The free-standing composites anodes via the facile one-step preparation showed high specific capacity, good rate capability, and long-term stability. Their good performance can be ascribed to the following reasons: i) the unique structure provided by the electrospinning can alleviate the volume change during the repeated charge/discharge; ii) the good conductivity given by the integration of carbon nanofiber and active metal selenides materials could promote fast electron transfer; iii) the absence of the binder, extra conductive materials, and current collector not only reduces the weight of the anode, but also eliminate the uncontrollable structures caused by slurry making. When extended the application to SIBs, the composites demonstrated great potentials. The postmortem analysis further confirmed the robust structure stability of the rationally designed composites. In conclusion, the free-standing metal selenides and carbon nanofibers composites are very promising electrode materials for next-generation flexible and lightweight batteries that are crucial for various emerging applications such as wearable electronics.

628642 Tuning Photocurrent Responses from Photosystem I Interfaced with Tailored Plasmonic Gold and Silver Nanopatterns

Abstract:

Photosystem I (PSI), a photosynthetic membrane protein, undergoes light-activated charge separation and unidirectional electron transfer with near-unity quantum efficiency. The robust photo-electrochemical (PEC) activities of PSI makes it an ideal candidate for bio-hybrid optoelectronic devices. But, the first step toward the fabrication of such devices requires its successful integration with suitable biotic-abiotic interfaces for the rational design of PSI-based PEC electrode assemblies. Herein, we present our fundamental investigations into tuning the optoelectronic properties of multi-chromophore PSI complexes by interfacing them with plasmonic noble metal nanopatterns tailored for specific plasmonic absorption wavelengths and hence, resonance modes. Our hypothesis is that PSI assembly with tailored plasmonic nanopatterns can provide a rational strategy to enhance the light harvesting properties and photon-to-electrochemical energy conversions in PSI thin films via electric field coupling with the localized surface plasmon resonance (LSPR) effects of the nanostructures. Such efforts can overcome the inherent limitations of low light absorption and poor quantum efficiencies in conventional PSI thin-film assemblies on inorganic substrates. Our earlier works had reported the first-ever experimental observation of plasmon-induced photocurrent enhancements from PSI immobilized on Fischer patterns of Ag-nanopyramids (Ag-NPs) whose plasmonic resonant peaks were tuned to the PSI absorption peaks at ~450 and ~680 nm.[1] Our recent investigations have extended these studies for spatially localized and spectrally resolved plasmon-enhanced photocurrents from PSIs assembled around highly ordered Au and Ag nano-discs (NDs) whose dipolar plasmon resonance modes are tuned to the wavelengths of ~680 nm and ~560 nm, respectively.[2] Specifically, our results indicate ~6.8 and ~17.5-fold enhancements in PSI photocurrents under the excitation wavelengths of ~680 and ~565 nm respectively, as compared to PSI assembled on planar ITO substrates. We also report the spectrally-resolved specificity of PSI-plasmon interactions via action spectra for photocurrent enhancements recorded over wavelength ranges of 395–810 nm using nine discrete LED light sources; while, spatially resolved photocurrent enhancements are estimated from FDTD simulations on Au and Ag ND patterns. The significant findings from these investigations include: 1) direct correlations between wavelength-dependent photocurrent enhancements from PSI assemblies and the plasmonic resonance modes for the respective nanopatterned substrates, and 2) broadband photocurrent enhancements from plasmon-coupled photo-activations of the otherwise blind chlorophyll regions of the native PSI absorption spectra.

References:

1. Pamu R, Sandireddy VP, Kalyanaraman R, Khomami B, Mukherjee D. Plasmon-Enhanced Photocurrent from Photosystem I Assembled on Ag Nanopyramids. *The Journal of Physical Chemistry Letters*. 2018; 9(5):970-977.
2. Pamu R, Lawrie BJ, Khomami B, Mukherjee D. Broadband Plasmonic Photocurrent Enhancement from Photosystem I Assembled with Tailored Arrays of Au and Ag Nanodisks. *ACS Applied Nano Materials*. 2021.

625374 High Charge Transfer and Charge Separation Donor-Electret-Acceptor Framework for Solar Cell Applications

Abstract:

Solar cells try to mimic the natural process of photosynthesis to convert the light energy into electrical energy. Long-range electron transfers occur rapidly over 10 Å distances between donor-acceptor pairs within the protein matrix. However, protein based moieties are not suited for the electronic applications. Widely employed Donor-(Bridge)$_n$-Acceptor (DB_nA; $n \geq 1$) framework typically fails to effectively transfer the charge over longer distances leading to larger exciton binding energies and lower power conversion efficiencies in organic solar cells. Modifying the dipole neutral bridge into an electret we propose a Donor-Electret$_n$-Acceptor (DE_nA; $n \geq 1$) model which addresses the limitations of conventional DBA architecture. (1) DEAs offer higher charge transfer (CT) and charge separation (CS) which overcomes the inherent length dependecy of DBAs with increasing chain length. (2) CS over ~12 Å and beyond is possible with the proposed DEA architecture which can compete with natural protein helix in efficient electron transfer and exciton dissociation. (3) Electret allows to tune the ground state (GS) dipole without comprimising on higher CT and CS properties; lower GS dipole enables the solubitliy of DEAs in milder solvents. (4) Higher degrees of freedom in the substitution patterns provided by the electret allows not only to alter CT and CS but also other electronic and optical properties like HOMO-LUMO gaps, optical gaps, dipoles, oscillator strengths etc. (5) Opens a new window in exploiting the substitution patterns provided by the electret to design novel materials which have potential applications in diverse areas. Our DFT/TD-DFT studies employing optimally tuned range separated hybrids predict accurate optical properties against the experimental results.

625641 First-Principles Study on the Phase Transition and Fast Ionic Transport in Lithium Spinel Anode Material

Abstract:

With the growing interest in power sources for large applications such as an electric vehicle (EV), lithium-ion batteries are finding new opportunities in this emerging area. Intensive research efforts are focused on developing suitable electrode material, the key component of lithium-ion batteries, for these applications. The electrode materials for lithium-ion batteries for use in EV require high stability, energy, and power. In this talk, I will present the first-principles understanding on the reaction mechanism of fast-charging lithium titanate anode material[1]. We reconcile the apparent contradiction between the high-power capability of LTO and the poor lithium-ion conductivity of its end-member phases. This insight presents new opportunities in searching for high-rate electrode materials.

References

[1] W. Zhang,‡ Dong-Hwa Seo,‡ T. Chen,‡ L. Wu, M. Topsakal, Y. Zhu, D. Lu, G. Ceder,* F. Wang*, "Kinetic Pathways of Ionic Transport in Fast Charging Lithium Titanate", Science, 367 (6481): 1030-1034 (2020)

624909 Structural Characterization of Carbon Composites Derived from Lignin Precursors for Electrodes in Batteries

Abstract:

Lignin, as an abundant natural polymer, has demonstrated great potential as a feedstock for carbon composites, which could be applied as electrodes in electrochemical devices. The lack of the knowledge of the effects of the lignin precursors on the carbon composites structures hampers wider adoption. In this work, different kinds of lignin, including yellow pine (YP, softwood lignin), hybrid poplar (HP, hardwood lignin) and switchgrass (SG, grass lignin), were adopted as the carbon precursors for comparison to elucidate the relationship between feedstock source and electrochemical performance of resulting carbon composites. Fourier-transform ion cyclotron resonance mass spectrometry (FT-ICR MS) combined with nuclear magnetic resonance (NMR) spectroscopy were conducted to obtain an unambiguous understanding of the chemical structures of lignin precursors. The produced carbon composites were tested as electrodes in lithium-ion batteries. The pair distribution functions (PDFs) of samples were collected from the neutron scattering diffraction at spallation neutron source in Oak Ridge National Laboratory. The changes in the PDFs corresponds to the samples from different precursors were compared with previous molecular dynamics simulation results.

628186 The Durability of Platinum Overlayers Formed By Self-Terminating Electrodeposition for PEM Fuel Cell Application

Abstract:

Proton exchange membrane fuel cells (PEMFCs) are an energy-efficient alternative to combustion engines for automotive applications. However, they are still on the cusp of mass adoption, mainly due to higher costs. Significant advances have been made in PEMFC system design over the last three decades concerning cost reduction and structural redesign. The membrane electrode assembly (MEA), especially the electrode, is considered 'the heart' of a PEMFC and is designed to accommodate the need for efficient transport of electrons, reactants, and heat as well as constraints imposed by the platinum cost. Advances in PEMFC electrode design over the years range from using platinum black films with a loading of 10 g_{pt}/cm^2 in the 1970s to present-day Pt/ PGM (Platinum Group Metal) nanoparticle coated carbon-black particles (Pt/C) that use about 0.3 mg_{pt}/cm^2. The electrocatalyst architecture based on highly dispersed nanoparticles on carbon black particles enables significant gain in specific surface area of the catalyst. However, durability problems associated with corrosion of carbon support and subsequent loss of active surface area during start-up and shut down cycles have led to renewed interest in carbon-free nanostructured electrodes that employ a thin coating of Platinum or PGM-based catalytic layer on mesostructured conductive support ((Debe 2012), (Liu et al. 2012))

In this context, we carried out systematic ex-situ investigations on the durability of platinum thin films comprising of atomic overlayers formed by Self-Terminating platinum electrodeposition (Liu et al. 2012). We characterized the sequential growth of platinum film using XPS and AFM measurements; and analyzed the electrocatalytic performance as a function of the number of self-terminating pulses. We found that using eight pulses of electrodeposition, corresponding to ~10 nm thick platinum film, can meet stringent ex-situ durability targets set for a potential electrocatalyst in Fuel Cell Vehicles (FCVs) application by the Department of Energy (DOE), USA. Our results also show that samples with platinum deposited using four pulses (< 4 $\mu g/cm^2$ loading) or even one pulse (< 1 $\mu g/cm^2$ loading) are electrochemically active. For eight and more pulses of platinum, we observed ~ 20% loss in Electrochemically Active Surface Area (ECSA) during the first 3000 durability cycles and 10-15% ECSA loss over the next 27000 cycles, indicating stabilization of the activated platinum morphology over time. In this presentation we will present results for the minimum number of pulses of platinum electrodeposition needed to form a complete overlayer of electrochemically active platinum and compare estimates of platinum loss using ICP-OES and Cyclic Voltammetry. These studies pave the way for an additive, roll-to-roll, and cost-effective manufacturing process of ultra-low platinum loaded MEAs for PEMFCs by building on our earlier results on printing silver nanostructures using a simple inkjet printer by *Print-Expose-Develop* technique (Parmar and Santhanam 2014) in conjunction with the ability to controllably form overlayers of platinum on metallic substrates by Self-terminating electrodeposition.

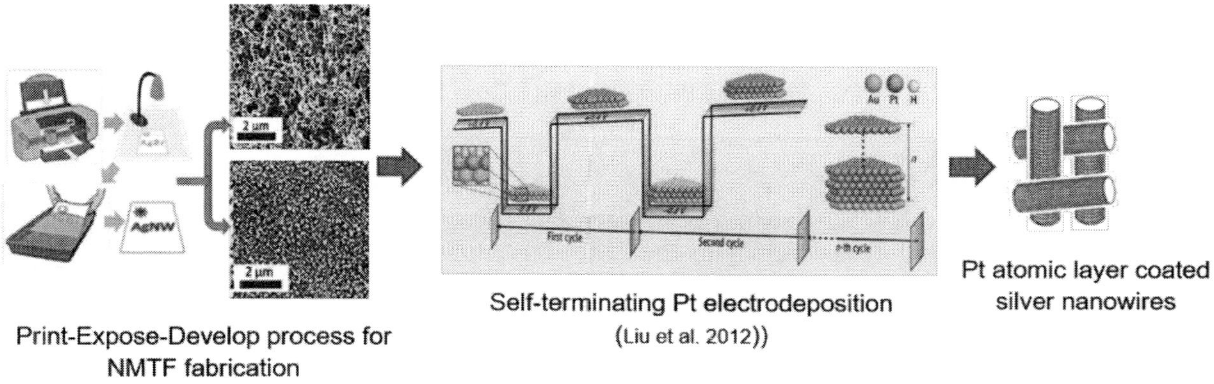

Fig.1 Schematic illustrating the process flow for forming electrocatalyst layers

631390 Optimized Continuous Millifluidic Surface Modification of Synthesized Silver Nanowires with Palladium

Abstract:

Silver nanowires (AgNWs) hold promise as the conductive material in conductive inks suitable for printing processes to create conductive patterns thanks to their three combined characteristics: high conductivity, transparency, and mechanical flexibility. The cost associated with raw material production for this application is among the most important factors in material selection and AgNWs can be synthesized simply in small scale batch process. However, rapid synthesis of high aspect ratio of AgNWs with high concentration is still a challenge that makes a huge gap in their large-scale practical applications. Moreover, a mixture of AgNWs and other silver nanostructures in organic solvent are obtained that need to be separated which makes the cost of purification and recovery of synthesized AgNWs a further barrier towards their practical applications. Application of a millifluidic reactor is a method to overcome the challenges in batch AgNWs synthesis through control of a uniform chemical and thermal environment in a small reaction volume to produce high aspect ratio and high yield of AgNW. Millifluidic reactors are an improvement over batch reactors, but optimization is still required to produce 100% yield of the desired morphology. In addition, AgNWs chemical stability is critical in practical applications as their oxidation can not only weaken mechanical stability but also increase the sheet resistance of transparent conductive films (TCFs). To eliminate or mitigate the oxidation concern, the surface of AgNWs can be alloyed or coated with graphene, graphene oxide, titanium oxide, zinc oxide, and noble metals such as gold and palladium (Pd). The formation of high reduction potential alloys on the surface of AgNWs is a promising technique to improve the performance and stability of AgNW-based TCFs. It has been shown that the chemical stability of AgNWs towards oxidation can be achieved with very small amount of Pd as the noble metal modifier, which is promising for improving the chemical stability of the synthesized AgNWs.

In this study, Design of Experiments (DOE) is used to optimize the millifluidic AgNW synthesis. The goal is to find the optimized conditions that produce close to 100% yield of synthesized AgNWs. The factors being examined are temperature, flowrate, and reagents concentration. Once the optimized reaction conditions are established, the AgNWs are synthesized using a millifluidic polyol process. The galvanic replacement reaction is then used to modify the surface of synthesized AgNWs. Based on previous studies, the desired amount of Pd that is galvanically replaced on a single AgNW is around 5%. Again, DOE is used to optimize the atomic weight percentage of Pd deposited on the AgNWs. The factors being examined are pH, flowrate, and reagents concentration. Once optimized conditions are set for both reactions, they are connected for continuous synthesis and surface modification. The outlet polytetrafluoroethylene (PTEF) tubing of the optimized AgNW polyol process is submerged in an ultrasonic bath to break up agglomeration and to cool down the suspension. Once the outlet PTEF tubing exits the ultrasonic bath, it is connected to the optimized galvanic surface modification millifluidic reactor via a cross junction. The AgNW suspension in the outlet PTEF tubing reacts with two other millifluidic tubes containing the Pd precursor solution ($Pd(NO_3)_2.2H_2O$ aqueous solution) and stabilizing solution (ascorbic acid and polyvinylpyrrolidone (PVP) aqueous solution). After the completion of the galvanic replacement reaction, the Pd treated AgNWs are washed and separated, and redispersed in deionized water for further Transmission Electron Microscopy (TEM), Scanning Electron Microscopy (SEM), Energy Dispersive X-Ray (EDX), and Mass Spectrometry (MS) characterizations to illustrate the formation of desired amount of PdNPs on the surface of AgNWs.

A novel low-cost synthesis of modified AgNWs with high chemical stability against oxidation ready to be coupled with conductive ink formulation and writing machine to create TCFs is illustrated, which is capable to be scaled up for large-scale industry-relevant demand production. This study is supported by National Science Foundation (NSF Award ID 1939018).

632045 Application of a Continuous Flow Millifluidic Reactor Towards Optimizing Manufacturing Throughput for Molybdenum Carbide Nanoparticles

Abstract:

Transition Metal Carbides (TMCs) has been extensively investigated for their use for multiple catalytic transformations, including CO_2 hydrogenation, hydrodeoxygenation and isomerization reactions. Due to the harsh conditions in which most TMCs are prepared, much effort has been focused on discovering new synthesis methodologies that yield nanostructured products with high surface area and enhanced activity which can be leveraged for catalysis. Recently, a scalable solution-phase synthesis of phase pure α-MoC_{1-x} nanoparticles by thermolytic decomposition of $Mo(CO)_6$ at mild reaction conditions and their superior catalytic properties was been demonstrated and adapted to continuous flow production in a millifluidic reactor. As well as providing a means to produce nanoparticles on an industrial scale, continuous flow platforms coupled with statistical Design of Experiment (DoE) methods can provide a powerful method for understanding the reaction parameter space for this synthetic route. This approach facilitates the rapid and efficient optimization of the reaction, with the target of maximizing throughput for scaling production. The heat and mass transport characteristics provided by millifluidic flow reactors allows for efficient control over reagents and parameters. This is essential for maximizing the throughput of nanoparticle products obtained from millifluidic reactors for scaling up nanofabrication. In this work, we demonstrate the use of statistical DoE in tandem with response surface methodology (RSM) for a parametric screening analysis to optimize the throughput of a molybdenum carbide nanoparticle synthesis utilizing a millifluidic flow reactor. A 2^4 full factorial design was implemented to evaluate four input variables (reaction temperature, flow rate, percentage of oleylamine, and precursor concentration) that carry statistically significant effects on three responses (throughout, residence time, and yield). RSM was then used to optimize the system based on the input goal of maximizing throughput. The three significant input variables identified by the screening were evaluated using a Doehlert matrix in order to investigate each significant variable at a higher number of levels based on their significance in effecting the throughput. Preliminary results show an optimized throughput of which translates to an increase by a factor of ~2.84 compared to the previously reported value (0.775 g hr^{-1}). This DoE screening analysis and throughput optimization of MoC_{1-x} synthesis opens the door to an increased feasibility for industrial scale-up.

AUTHOR INDEX

AIChE
120 Wall Street, FL 23
New York, NY 10005-4020

ISBN 978-1-7138-5715-0

CURRAN ASSOCIATES INC.
proceedings.com